高等学校计算机应用规划教材

计算机网络实验
实训教程

王鸿运　王学春　主　编
郑晶晶　韩　芳　副主编

清华大学出版社
北　京

内容简介

《计算机网络实验实训教程》旨在帮助读者学习计算机网络基础知识之后，进行网络设备操作、配置、设计和综合应用的上机实战训练。内容涵盖了常用交换路由硬件设备知识和交换机与路由器的详细配置。每个实验均给出实验目的、实验要求、实验环境、实验步骤等，并且都安排了一定数量的上机实战操作，能够让读者通过实验操作和实战练习来巩固相关的知识并掌握技能。

本书内容翔实，具有很强的实用性，着重动手能力的培养。书中每一个实训都经过精心挑选，并在操作步骤上给予了详细说明。

本书可作为本专科院校计算机类专业、电子信息类专业、通信类专业、电子商务类专业及其他专业的计算机网络课程配套实验实训教材，也可作为大中专院校、成人高校、计算机培训学校等各类院校的实验实训指导用书。

本书对应的习题答案可以到 http://www.tupwk.com.cn/downpage 网站下载。

本书封面贴有清华大学出版社防伪标签，无标签者不得销售。

版权所有，侵权必究。举报：010-62782989，beiqinquan@tup.tsinghua.edu.cn。

图书在版编目(CIP)数据

计算机网络实验实训教程 / 王鸿运，王学春 主编. —北京：清华大学出版社，2015（2023.8重印）
(高等学校计算机应用规划教材)
ISBN 978-7-302-39888-2

Ⅰ. ①计… Ⅱ. ①王… ②王… Ⅲ. ①计算机网络—实验—高等学校—教材 Ⅳ. ①TP393-33

中国版本图书馆 CIP 数据核字(2015)第 080709 号

责任编辑：胡辰浩
装帧设计：孔祥峰
责任校对：邱晓玉
责任印制：宋 林

出版发行：清华大学出版社
网　　址：http://www.tup.com.cn，http://www.wqbook.com
地　　址：北京清华大学学研大厦 A 座　　邮　　编：100084
社 总 机：010-83470000　　邮　　购：010-62786544
投稿与读者服务：010-62776969，c-service@tup.tsinghua.edu.cn
质 量 反 馈：010-62772015，zhiliang@tup.tsinghua.edu.cn
课 件 下 载：http://www.tup.com.cn，010-62794504

印 装 者：三河市龙大印装有限公司
经　　销：全国新华书店
开　　本：185mm×260mm　　印　张：16.5　　字　数：381 千字
版　　次：2015 年 5 月第 1 版　　印　次：2023 年 8 月第 12 次印刷
定　　价：79.00 元

产品编号：064126-05

前　　言

在当今信息社会，随着 Internet 的深入普及，计算机网络应用几乎遍及人类活动的各个领域，计算机网络技术已被誉为是"近代最深刻的技术革命"，人们用"网络时代"和"网络经济"等术语来描述计算机网络对社会信息化与经济发展的影响。社会的信息化、数据的分布式处理、各种计算机资源的共享等应用需求，推动着计算机网络的迅速发展。

计算机网络是计算机技术与通信技术密切结合的学科，也是计算机应用中一个空前活跃的领域。该课程不仅是计算机科学与技术专业的主干课程，也是电子与通信专业学生以及广大从事计算机应用和信息管理的科技人员都必须学习的课程。同时，我国的信息化建设也需要大量掌握计算机网络基础知识和应用技术的专业人才。该课程不仅是一门理论性很强的课程，同时也是一门实践性很强的课程。因此，为了配合计算机网络的实验教学，以《计算机网络技术》为背景，结合实验室的设备和仪器编写了本实验指导书，以帮助学生通过严格的实践训练真正掌握和深入理解计算机网络的基本理论、协议和算法。

本书是面向理工科的本专科学生的计算机网络实验指导教材，结合理论教学精心设计了交换机实验 17 个、路由器实验 15 个、综合实训项目 11 个和综合性实验 5 个，有利于学生在掌握了一定的理论和实践之后进行设计和创新。

本书在实验内容组织上具有较强的系统性和可操作性，所要求的实验环境相对简单和单一，所有实验内容均在实验室完成。学生通过完成设计的实验内容能够深入掌握和理解计算机网络的工作原理和工作过程，增强处理实际问题的能力。

本课程实验教学要求学生完成交换机全部实验，教师可根据不同的专业方向选择交换路由全部实验，如有必要可对加强综合性实验进行整合实训，在附录中制作了 5 个代表性综合性实验为实训考核学生能力设计，一方面能够把所学的内容进行融合，另一方面可以充分发挥学生的创造性，把所学的知识运用到实践中。在课程组织上，可按以下方式进行。

(1) 学生每两人 1 小组，每小组两台计算机，由两台计算机组建成服务器和客户机，网络可按照需求进行配置实验，最终可以实现访问。

(2) 实验所需的工具、软件、材料基本为每组分配一套，而有些工具有多个，为各组共用，如网线测试仪等可以公用。

本书可作为本专科院校计算机类专业、电子信息类专业、通信类专业、电子商务类专业及其他专业的计算机网络课程配套实验实训教材，也可作为大中专院校、成人高校、计算机培训学校等各类院校的实验实训指导用书。

本书是由黄河科技学院教师王鸿运、王学春任主编，郑晶晶、韩芳任副主编，由于设备不断更新，实验实训的经验也在不断更新和完善，本书难免有不足之处，欢迎广大读者批评指正。我们的电话是 010-62796045，信箱是 huchenhao@263.net。

本书对应的习题答案可以到 http://www.tupwk.com.cn/downpage 网站下载。

作　者

2015 年 3 月

目 录

第一章 交换机实验 ·········· 1

实验一 交换机带外管理 ·········· 1
 一、实验目的 ·········· 1
 二、应用环境 ·········· 1
 三、实验设备 ·········· 1
 四、实验拓扑 ·········· 1
 五、实验要求 ·········· 2
 六、实验步骤 ·········· 2
 七、课后练习 ·········· 5
 八、相关配置命令详解 ·········· 6

实验二 交换机的配置模式 ·········· 7
 一、实验目的 ·········· 7
 二、应用环境 ·········· 7
 三、实验设备 ·········· 7
 四、实验拓扑 ·········· 7
 五、实验要求 ·········· 8
 六、实验步骤 ·········· 8
 七、注意事项和排错 ·········· 12
 八、思考题 ·········· 13
 九、课后练习 ·········· 13
 十、相关配置命令详解 ·········· 13

实验三 交换机 CLI 界面调试技巧 ·········· 16
 一、实验目的 ·········· 16
 二、应用环境 ·········· 17
 三、实验设备 ·········· 17
 四、实验拓扑 ·········· 17
 五、实验要求 ·········· 17
 六、实验步骤 ·········· 18
 七、课后练习 ·········· 20
 八、相关配置命令详解 ·········· 20

实验四 交换机恢复出厂设置及其基本配置 ·········· 23
 一、实验目的 ·········· 23
 二、应用环境 ·········· 23
 三、实验设备 ·········· 23
 四、实验拓扑 ·········· 24
 五、实验要求 ·········· 24
 六、实验步骤 ·········· 24
 七、注意事项和排错 ·········· 27
 八、配置序列 ·········· 27
 九、思考题 ·········· 27
 十、课后练习 ·········· 27
 十一、相关配置命令详解 ·········· 28

实验五 使用 Telnet 方式管理交换机 ·········· 30
 一、实验目的 ·········· 30
 二、应用环境 ·········· 30
 三、实验设备 ·········· 31
 四、实验拓扑 ·········· 31
 五、实验要求 ·········· 31
 六、实验步骤 ·········· 31
 七、注意事项和排错 ·········· 35
 八、配置序列 ·········· 35
 九、思考题 ·········· 35
 十、课后练习 ·········· 35
 十一、相关配置命令详解 ·········· 36

实验六 使用 Web 方式管理交换机 ·········· 40
 一、实验目的 ·········· 40
 二、应用环境 ·········· 40
 三、实验设备 ·········· 40
 四、实验拓扑 ·········· 40

| 五、实验要求……………………40 | 八、配置序列……………………61 |

　　六、实验步骤……………………41　　　　　　九、思考题………………………63
　　七、注意事项和排错……………42　　　　　　十、相关配置命令详解…………63
　　八、配置序列……………………43　　实训二　跨交换机相同 VLAN
　　九、思考题………………………43　　　　　　间通信……………………65
　　十、课后练习……………………43　　　　　一、实训设备……………………65
　　十一、相关配置命令详解………43　　　　　二、实训拓扑……………………65
实训一　Telnet 和 Web 方式　　　　　　实验九　交换机 MAC 与 IP 的
　　　　　管理交换机………………44　　　　　　绑定………………………68
　　一、实验目的……………………44　　　　　一、实验目的……………………68
　　二、实验设备……………………44　　　　　二、应用环境……………………69
　　三、实验拓扑……………………45　　　　　三、实验设备……………………69
　　四、实验内容……………………45　　　　　四、实验拓扑……………………69
　　五、实验步骤……………………45　　　　　五、实验要求……………………69
　　六、注意事项和排错……………48　　　　　六、实验步骤……………………70
　　七、课后练习……………………49　　　　　七、注意事项和排错……………71
实验七　交换机 VLAN 划分实验…49　　　　　八、相关配置命令详解…………71
　　一、实验目的……………………49　　实验十　生成树实验………………74
　　二、应用环境……………………49　　　　　一、实验目的……………………74
　　三、实验设备……………………50　　　　　二、应用环境……………………74
　　四、实验拓扑……………………50　　　　　三、实验设备……………………74
　　五、实验要求……………………50　　　　　四、实验拓扑……………………74
　　六、实验步骤……………………51　　　　　五、实验要求……………………75
　　七、注意事项和排错……………53　　　　　六、实验步骤……………………75
　　八、配置序列……………………53　　　　　七、注意事项和排错……………77
　　九、思考题………………………55　　　　　八、配置序列……………………78
　　十、课后练习……………………55　　　　　九、课后练习……………………78
　　十一、相关配置命令详解………55　　　　　十、思考题………………………78
实验八　跨交换机相同 VLAN　　　　　　　　十一、相关配置命令详解………78
　　　　　间通信……………………57　　实验十一　交换机链路聚合………78
　　一、实验目的……………………57　　　　　一、实验目的……………………78
　　二、应用环境……………………57　　　　　二、应用环境……………………79
　　三、实验设备……………………57　　　　　三、实验设备……………………79
　　四、实验拓扑……………………57　　　　　四、实验拓扑……………………79
　　五、实验要求……………………58　　　　　五、实验要求……………………80
　　六、实验步骤……………………58　　　　　六、实验步骤……………………80
　　七、注意事项和排错……………61　　　　　七、注意事项和排错……………83

八、课后练习 ················84
　　九、相关配置命令详解 ········84
实训三　基于有VLAN的
　　　　交换机链路聚合 ········88
　　一、实训目的 ················88
　　二、实训拓扑 ················88
　　三、实训要求 ················88
　　四、实训步骤 ················89
实验十二　认识三层交换机 ········90
　　一、实验目的 ················90
　　二、应用环境 ················90
　　三、实验设备 ················90
　　四、实验拓扑 ················91
　　五、实验要求 ················91
　　六、实验步骤 ················91
　　七、注意事项和排错 ··········94
　　八、相关配置命令详解 ········94
　　九、课后练习 ················98
实验十三　多层交换机VLAN的划
　　　　　分和VLAN间路由 ······99
　　一、实验目的 ················99
　　二、应用环境 ················99
　　三、实验设备 ················99
　　四、实验拓扑 ················99
　　五、实验要求 ···············100
　　六、实验步骤 ···············100
　　七、注意事项和排错 ·········103
　　八、配置序列 ···············103
　　九、思考题 ·················106
　　十、课后练习 ···············106
　　十一、相关配置命令详解 ·····106
实验十四　多层交换机实现二层
　　　　　交换机 VLAN 之间
　　　　　路由 ···············109
　　一、实验目的 ···············109
　　二、应用环境 ···············110
　　三、实验设备 ···············110

　　四、实验拓扑 ···············110
　　五、实验要求 ···············110
　　六、实验步骤 ···············111
　　七、注意事项和排错 ·········115
　　八、思考题 ·················115
　　九、相关配置命令详解 ·······115
实训四　三层交换机实现二层
　　　　交换机不同VLAN
　　　　间通信 ···············118
　　一、实训拓扑 ···············118
　　二、实训要求 ···············119
　　三、实训步骤 ···············119
实验十五　多层交换机静态
　　　　　路由实验 ···········121
　　一、实验目的 ···············121
　　二、应用环境 ···············121
　　三、实验设备 ···············121
　　四、实验拓扑 ···············121
　　五、实验要求 ···············122
　　六、实验步骤 ···············122
　　七、注意事项和排错 ·········127
　　八、思考题 ·················127
实训五　多层交换机的静态
　　　　路由配置 ·············128
　　一、实训设备 ···············128
　　二、实训拓扑 ···············128
　　三、实训要求 ···············128
　　四、实训步骤 ···············128
实验十六　三层交换机 RIP
　　　　　动态路由 ···········129
　　一、实验目的 ···············129
　　二、应用环境 ···············129
　　三、实验设备 ···············129
　　四、实验拓扑 ···············130
　　五、实验要求 ···············130
　　六、实验步骤 ···············131
　　七、注意事项和排错 ·········137

八、思考题·················137
　　九、相关配置命令详解·········137
实训六　多层交换机的动态 rip
　　　　路由配置···············145
　　一、实训设备···············145
　　二、实训拓扑···············146
　　三、实训要求···············146
　　四、实训步骤···············146
实验十七　三层交换机 OSPF
　　　　动态路由···············148
　　一、实验目的···············148
　　二、应用环境···············148
　　三、实验设备···············148
　　四、实验拓扑···············148
　　五、实验要求···············149
　　六、实验步骤···············149
　　七、注意事项和排错·········154
　　八、思考题·················155
实训七　多层交换机之间的
　　　　动态 ospf 配置··········163
　　一、实训设备···············163
　　二、实训拓扑···············163
　　三、实训要求···············164
　　四、实训步骤···············164

第二章　路由器实验·············166
实验一　路由器接口简介·········166
　　一、实验目的···············166
　　二、应用环境···············166
　　三、实验设备···············166
　　四、实验拓扑···············166
　　五、实验要求···············167
　　六、实验步骤···············167
　　七、注意事项和排错·········167
实验二　路由器的基本管理方法···167
　　一、实验目的···············167
　　二、应用环境···············168

　　三、实验设备···············168
　　四、实验拓扑···············168
　　五、实验要求···············168
　　六、实验步骤···············168
　　七、注意事项和排错·········173
　　八、思考题·················173
　　九、课后练习···············174
实验三　路由器的基本配置·······174
　　一、实验目的···············174
　　二、应用环境···············174
　　三、实验设备···············174
　　四、实验拓扑···············174
　　五、实验要求···············175
　　六、实验步骤···············175
　　七、注意事项和排错·········180
　　八、配置序列···············180
　　九、思考题·················180
　　十、课后练习···············180
　　十一、相关命令详解·········180
实验四　路由器的文件维护·······184
　　一、实验目的···············184
　　二、应用环境···············184
　　三、实验设备···············184
　　四、实验拓扑···············184
　　五、实验要求···············184
　　六、实验步骤···············185
　　七、注意事项和排错·········187
　　八、配置序列···············188
　　九、思考题·················188
　　十、课后练习···············188
实验五　单臂路由实验···········188
　　一、实验目的···············188
　　二、应用环境···············188
　　三、实验设备···············188
　　四、实验拓扑···············188
　　五、实验要求···············188
　　六、重要配置···············189

实验六　路由器静态路由配置……189
一、实验目的……189
二、应用环境……190
三、实验设备……190
四、实验拓扑……190
五、实验要求……190
六、实验步骤……190
七、注意事项和排错……193
八、配置序列……193
九、思考题……194

实训八　多台路由器之间静态路由配置……194
一、实训设备……194
二、实训拓扑……194
三、实训要求及步骤……194

实验七　路由器 RIP-1 配置……195
一、实验目的……195
二、应用环境……195
三、实验设备……196
四、实验拓扑……196
五、实验要求……196
六、实验步骤……196
七、注意事项和排错……200
八、配置序列……200
九、思考题……201

实训九　多台路由器的动态 RIP-1 路由配置……201
一、实训设备……201
二、实训拓扑……201
三、实训要求及步骤……202

实验八　路由器 RIP-2 配置……203
一、实验目的……203
二、应用环境……203
三、实验设备……203
四、实验拓扑……203
五、实验要求……203
六、实验步骤……204
七、注意事项和排错……207
八、配置序列……207
九、思考题……208
十、相关命令详解……208

实训十　多台路由器的动态 RIP-2 路由配置……208
一、实训设备……208
二、实训拓扑……209
三、实训要求及步骤……209

实验九　静态路由和直连路由引入配置……210
一、实验目的……210
二、应用环境……210
三、实验设备……210
四、实验拓扑……210
五、实验要求……210
六、实验步骤……211
七、注意事项和排错……212
八、配置序列……213
九、思考题……214
十、课后练习……214
十一、相关命令详解……214

实验十　单区域 OSPF 基本配置……216
一、实验目的……216
二、应用环境……216
三、实验设备……216
四、实验拓扑……216
五、实验要求……216
六、实验步骤……217
七、注意事项和排错……220
八、配置序列……220
九、思考题……221
十、课后练习……221
十一、相关命令详解……221

实验十一　RIP-2 邻居认证配置……222
一、实验目的……222
二、应用环境……222

三、实验设备……………………222
　　四、实验拓扑……………………222
　　五、实验要求……………………222
　　六、实验步骤……………………223
　　七、注意事项和排错……………225
　　八、配置序列……………………225
　　九、思考题………………………226
　　十、相关命令详解………………226
实验十二　多区域 OSPF 配置……227
　　一、实验目的……………………227
　　二、应用环境……………………227
　　三、实验设备……………………227
　　四、实验拓扑……………………227
　　五、实验要求……………………227
　　六、实验步骤……………………227
　　七、注意事项和排错……………229
　　八、配置序列……………………229
　　九、思考题………………………230
　　十、相关命令详解………………230
实训十一　多台路由器间的动态
　　　　　OSPF 路由配置…………231
　　一、实训设备……………………231
　　二、实训拓扑……………………231
　　三、实训要求及步骤……………231
实验十三　OSPF 邻居认证配置……232
　　一、实验目的……………………232
　　二、应用环境……………………232
　　三、实验设备……………………232
　　四、实验拓扑……………………233
　　五、实验要求……………………233
　　六、实验步骤……………………233
　　七、注意事项和排错……………235
　　八、配置序列……………………235

　　九、思考题………………………235
　　十、课后练习……………………235
　　十一、相关命令详解……………235
实验十四　OSPF 路由汇总配置……236
　　一、实验目的……………………236
　　二、应用环境……………………236
　　三、实验设备……………………236
　　四、实验拓扑……………………236
　　五、实验要求……………………236
　　六、实验步骤……………………237
　　七、注意事项和排错……………239
　　八、配置序列……………………239
　　九、思考题………………………239
　　十、课后练习……………………240
　　十一、相关命令详解……………240
实验十五　NAT 地址转换的
　　　　　配置………………………241
　　一、实验目的……………………241
　　二、应用环境……………………241
　　三、实验设备……………………241
　　四、实验拓扑……………………241
　　五、实验要求……………………241
　　六、实验步骤……………………242
　　七、注意事项和排错……………243
　　八、配置序列……………………243
　　九、思考题………………………244
　　十、课后练习……………………245
　　十一、相关命令详解……………245
附录………………………………………247
参考文献…………………………………254

第一章 交换机实验

实验一 交换机带外管理

一、实验目的

1. 熟悉普通二层交换机的外观。
2. 了解普通二层交换机各端口的名称和作用。
3. 了解交换机最基本的管理方式——带外管理的方法。

二、应用环境

网络设备的管理方式可以简单地分为带外管理(out-of-band)和带内管理(in-band)两种管理模式。所谓带内管理，是指网络的管理控制信息与用户网络的承载业务信息通过同一个逻辑信道传送，简而言之，就是占用业务带宽；而在带外管理模式中，网络的管理控制信息与用户网络的承载业务信息在不同的逻辑信道传送，也就是设备提供专门用于管理的带宽。

目前很多高端的交换机都带有带外网管接口，使网络管理的带宽和业务带宽完全隔离，互不影响，构成单独的网管网。

通过Console口管理是最常用的带外管理方式，通常用户会在首次配置交换机或者无法进行带内管理时使用带外管理方式。

带外管理方式也是使用频率最高的管理方式。带外管理时，我们可以采用Windows操作系统自带的超级终端程序来连接交换机，当然，用户也可以采用自己熟悉的终端程序。

Console口：也叫配置口，用于接入交换机内部对交换机进行配置。

Console线：交换机包装箱中标配线缆，用于连接Console口和配置终端。

三、实验设备

1. DCS-3926S交换机   1台。
2. PC机   1台。
3. 交换机Console线   1根。

四、实验拓扑

该实验拓扑结构如图1-1所示。

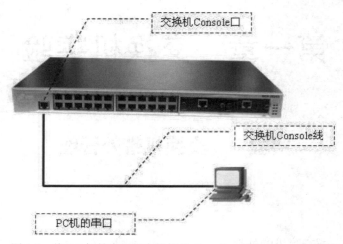

图 1-1　将 PC 机的串口和交换机的 Console 口用 Console 线连接

五、实验要求

1. 正确认识交换机上各端口名称。
2. 熟练掌握使用交换机Console线连接交换机的Console口和PC的串口。
3. 熟练掌握使用超级终端进入交换机的配置界面。

六、实验步骤

第一步：认识交换机的端口。

交换机的端口外观如图 1-2 所示，"0/0/1"中的第一个"0"表示堆叠中的第一台交换机，如果是"1"，就表示第 2 台交换机；第 2 个"0"表示交换机上的第 1 个模块(DCS-3926s 交换机有 3 个模块：网络端口模块 0(M0)，模块 1(M1)，模块 2(M2)；最后的"1"表示当前模块上的第 1 个网络端口)。

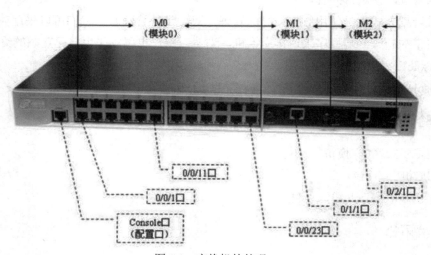

图 1-2　交换机的外观

"0/0/1"表示用户使用的是堆叠中第一台交换机网络端口模块上的第一个网络端口。默认情况下，如果不存在堆叠，交换机总会认为自己是第 0 台交换机。

第二步：连接 Console 线。

插拔 Console 线时注意保护交换机的 Console 口和 PC 的串口，不要带电拔插。

第三步：使用超级终端连入交换机。

(1) 打开 Windows 系统，选择"开始"|"程序"|"附件"|"通讯"|"超级终端"命令，将打开"连接描述"对话框，如图 1-3 所示。在"名称"行输入"DCS-3926S"，表示新建连接的名称，系统会为用户把这个连接保存在附件中的通讯栏中，以便用户的下次使用。然后单击"确定"按钮。

图 1-3　"连接描述"对话框

(2) 选择所使用的端口号：第一行的"DCS-3926S"是上一个对话框中填入的"名称"，最后一行的"连接时使用"下拉列表框默认设置是连接在"COM1"口上，其他选项视用户实际连接的端口而定，如图 1-4 所示。

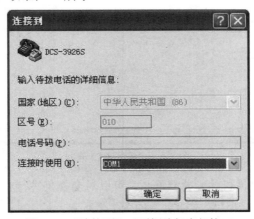

图 1-4　"连接到"对话框选择串行接口

(3) 通过"COM1 属性"对话框设置端口属性，如图 1-5 所示。单击右下方的"还原默认值"按钮，"每秒位数"设置为 9600，"数据位"设置为 8，"奇偶校验"设置为无，"停止位"设置为 1，"数据流控制"设置为无。

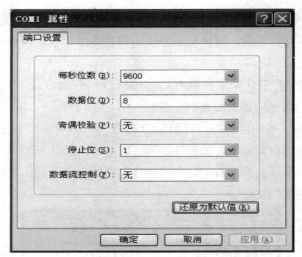

图 1-5 设置串口属性

(4) 如果 PC 机串口与交换机的 Console 口连接正确，那么在超级终端中按下 Enter 键，将会看到如图 1-6 所示的界面，表示已经进入了交换机，此时就可以对交换机输入指令进行查看。

图 1-6 交换机 CLI 界面

(5) 用户成功进入交换机的配置界面后，可以对交换机进行必要的配置。Show version 命令可以查看交换机的软硬件版本信息，如图 1-7 所示。

(6) 使用show running查看当前配置。

Switch>enable   !进入特权配置模式(详见实验 2)
switch#show running-config

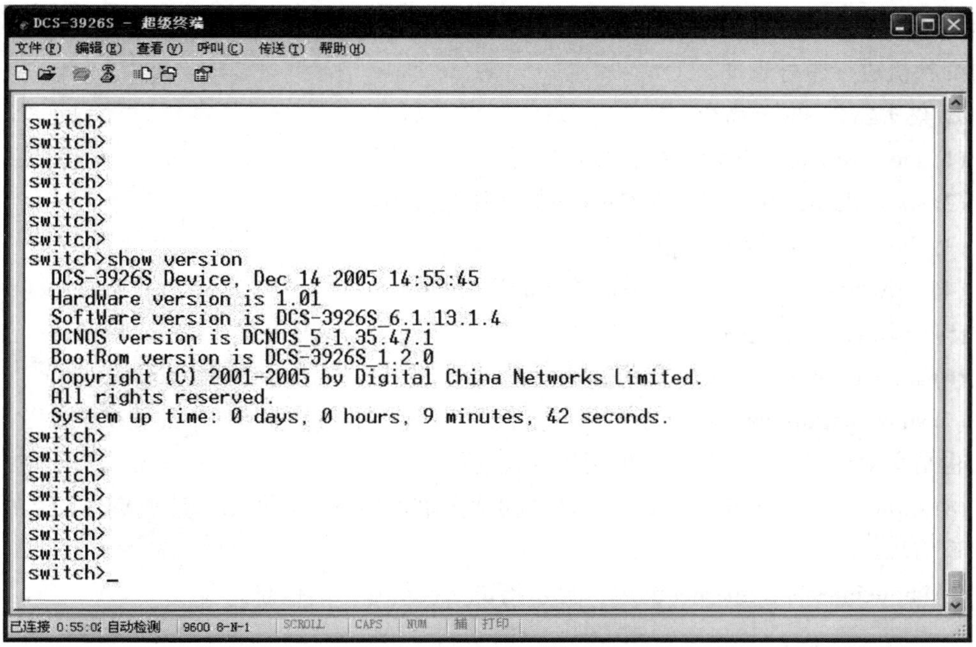

图 1-7　交换机硬件版本信息

Current configuration:
!
　　hostname switch
!
Vlan 1
!
Interface Ethernet0/0/1
!
Interface Ethernet0/0/2
!
Interface Ethernet0/0/3
!
Interface Ethernet0/0/4
!
……
Interface Ethernet0/0/23
!
Interface Ethernet0/0/24
!
!
switch#

七、课后练习

1. 如果用户的笔记本电脑上没有能连接 Console 线的串口，那么可以在电脑配件市场上购买一根 USB 转串口的线缆，在自己的电脑上安装该线缆的驱动程序，使用电脑的 USB 口对交换机进行带外管理。

2. 熟悉常用 show 命令。

(1) show version：显示交换机版本信息。

(2) show flash：显示保存在 flash 中的文件及大小。

(3) show arp：显示 ARP 映射表。

(4) show history：显示用户最近输入的历史命令。

(5) show rom：显示启动文件及大小。

(6) show running-config：显示当前运行状态下生效的交换机参数配置。

(7) show startup-config：显示当前运行状态下写在 Flash Memory 中的交换机参数配置，通常也是交换机下次上电启动时所用的配置文件。

(8) show switchport interface：显示交换机端口的VLAN端口模式和所属VLAN号及交换机端口信息。

(9) show interface ethernet 0/0/1：显示指定交换机端口的信息。

八、相关配置命令详解

show running-config

命令：show running-config。

功能：显示当前运行状态下生效的交换机参数配置。

默认情况：对于正在生效的配置参数，如果与默认工作参数相同，则不显示。

命令模式：特权用户配置模式。

使用指南：当用户完成一组配置后，需要验证是否配置正确，则可以执行 show running-config 命令来查看当前生效的参数。

举例：

Switch#show running-config

show version

命令：show version。

功能：显示交换机版本信息。

命令模式：特权用户配置模式。

使用指南：通过查看版本信息可以获知硬件和软件所支持的功能特性。

举例：

Switch#show version
DCS-3926S Device, Mar 17 2004 11:18:20
HardWare version is 2.00, SoftWare version is RW-0.0.47, BootRom versionis 1.1.6
Copyright (C) 2001-2002 by Digitalchina Networks Limited.
All rights reserved.

实验二 交换机的配置模式

一、实验目的

1. 了解交换机不同配置模式的功能。
2. 了解交换机不同配置模式的进入和退出方法。

二、应用环境

在实验一中，我们可以成功地进入交换机的配置界面。我们所看到的配置界面称之为 CLI 界面。CLI 界面又称为命令行界面，和图形界面(GUI)相对应。CLI 的全称是 Command Line Interface，它由 Shell 程序提供，由一系列的配置命令组成，根据这些命令在配置管理交换机时所起的作用不同，Shell 将这些命令分类，不同类别的命令对应着不同的配置模式。

命令行界面是交换机调试界面中的主流界面，基本上所有的网络设备都支持命令行界面。国内外主流的网络设备供应商使用很相近的命令行界面，方便用户调试不同厂商的设备。神州数码网络产品的调试界面兼容国内外主流厂商的界面，和思科命令行接近，便于用户学习。只有少部分厂商使用自己独有的配置命令。

三、实验设备

1. DCS-3926S 交换机   1 台。
2. PC 机   1 台。
3. Console 线   1 根。

四、实验拓扑

该实验拓扑结构如图 1-8 所示。

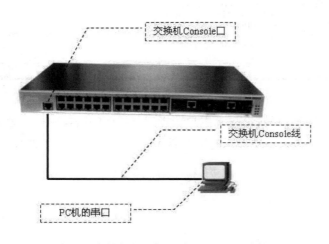

图 1-8 交换机与主机通过 Console 口连接图

五、实验要求

1. 熟悉 Setup 配置模式。
2. 熟悉一般用户配置模式。
3. 熟悉特权用户配置模式。
4. 了解全局配置模式。
5. 了解接口配置模式。
6. 了解 VLAN 配置模式。

六、实验步骤

第一步：Setup 模式的配置方法。

交换机出厂第一次启动，进入"Setup Configuration"，如图 1-9 所示。用户可以选择进入 Setup 模式或者跳过 Setup 模式。在此界面下通过键盘输入"y"，再按 Enter 键就会进入到 Setup 模式。

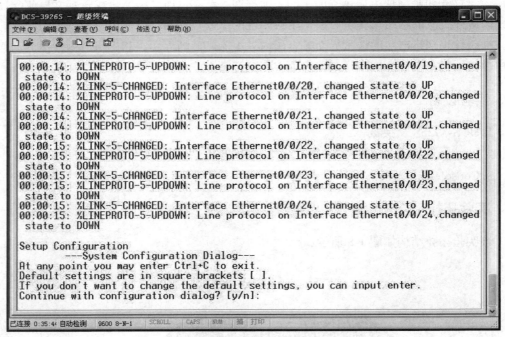

图 1-9　交换机启动界面

在进入主菜单之前，系统会提示用户选择配置界面的语言种类，对英文不是很熟悉的用户可以选择"1"，进入中文提示的配置界面，如图 1-10 所示。选择"0"则进入英文提示的配置界面。

Please select language
[0]:English
[1]:中文
Selection(0|1)[0]:

图 1-10 "Setup Configuration" 中文界面

下面是 Setup 主菜单的提示：

Configure menu	！配置菜单
[0]:Config hostname	！配置交换机的名字
[1]:Config interface-Vlan1	！配置交换机的管理 IP
[2]:Config telnet-server	
[3]:Config web-server	
[4]:Config SNMP	
[5]:Exit setup configuration without saving	！不保留配置，退出 Setup 配置模式
[6]:Exit setup configuration after saving	！保留配置，退出 Setup 配置模式
Selection number:	

在 Setup 主菜单上选择序号 "5"，退出 Setup 配置模式的同时用户在 Setup 模式下所做的配置均不保留。

在 Setup 主菜单上选择序号 "6"，退出 Setup 配置模式的同时用户在 Setup 模式下所做的配置均保留。如用户在 Setup 配置模式下，设置了 IP 地址、打开了 Web 服务，选择序号 "6" 退出 Setup 主菜单后，用户就可以通过 PC 对交换机进行 HTTP 管理配置。

第二步：一般用户配置模式的配置方法。

退出 Setup 模式即进入一般用户配置模式，也可以称为 ">" 模式。该模式的命令比较少，通过 "?" 命令使用帮助，如图 1-11 所示。

说明在该模式下，只有 enable、exit、help、show 这 4 个命令可以使用。

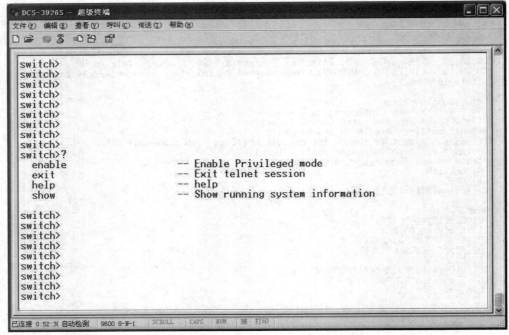

图1-11 通过"?"命令使用帮助

第三步：特权用户配置模式的配置方法。

在一般用户配置模式下输入"enable"，进入特权用户配置模式。特权用户配置模式的提示符为"#"，所以也称为"#"模式，该模式界面如图1-12所示。

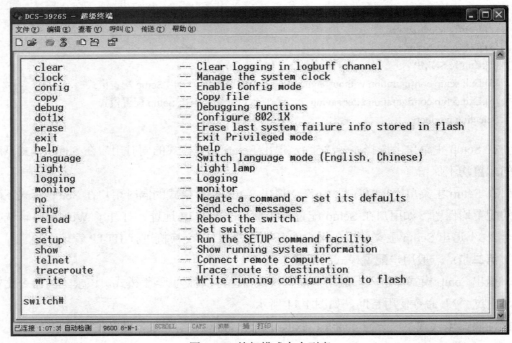

图1-12 特权模式命令列表

在特权用户配置模式下，用户可以查询交换机配置信息、各个端口的连接情况、收发数据统计等。而且进入特权用户配置模式后，可以进入到全局模式对交换机的各项配置进行修改，因此进行特权用户配置模式必须要设置特权用户口令，防止非特权用户的非法使用。

第四步：全局配置模式的配置方法。

在特权模式下输入"config terminal"或者"config t"或者"config"就可以进入全局配置模式。全局配置模式也称为"config"模式。

```
switch#config terminal
switch(config)#
```

在全局配置模式下，用户可以对交换机进行全局性的配置，如对 MAC 地址表、端口镜像、创建 VLAN、启动 IGMP Snooping、GVRP、STP 等功能进行配置。用户在全局模式式还可通过命令进入到端口对各个端口进行配置。

下面在全局配置模式下设置特权用户口令：

```
switch>enable
switch#config terminal                    ！进入全局配置模式，见第四步
switch(Config)#enable password level admin
Current password:                         ！原密码为空，直接回车
New password:*****
Confirm new password:*****                ！输入密码
switch(Config)#exit
switch#write
switch#
```

验证配置：

验证方法 1：重新进入交换机。

```
switch#exit                               ！退出特权用户配置模式
switch>
switch>enable                             ！进入特权用户配置模式
Password:*****
switch#
```

验证方法 2：show 命令来查看。

```
switch#show running-config
Current configuration:
enable password level admin 827ccb0eea8a706c4c34a16891f84e7b
  ！该行显示了已经为交换机配置了 enable 密码
    hostname switch
    Vlan 1
……                                        ！省略部分显示
```

第五步：接口配置模式的配置方法。

switch(Config)#interface ethernet 0/0/1
switch(Config-Ethernet0/0/1)#   ！已经进入以太端口 0/0/1 的接口
switch(Config)#interface vlan 1
switch(Config-If-Vlan1)#   ！已经进入 VLAN1 的接口，也就是 CPU 的接口

第六步：VLAN 配置模式的配置方法。

switch(Config)#vlan 100
switch(Config-Vlan100)#

验证配置：

switch(Config-Vlan100)#exit
switch(Config)#exit
switch#show vlan

VLAN	Name	Type	Media	Ports	
1	default	Static	ENET	Ethernet0/0/1	Ethernet0/0/2
				Ethernet0/0/3	Ethernet0/0/4
				Ethernet0/0/5	Ethernet0/0/6
				Ethernet0/0/7	Ethernet0/0/8
				Ethernet0/0/9	Ethernet0/0/10
				Ethernet0/0/11	Ethernet0/0/12
				Ethernet0/0/13	Ethernet0/0/14
				Ethernet0/0/15	Ethernet0/0/16
				Ethernet0/0/17	Ethernet0/0/18
				Ethernet0/0/19	Ethernet0/0/20
				Ethernet0/0/21	Ethernet0/0/22
				Ethernet0/0/23	Ethernet0/0/24
100	VLAN0100	Static	ENET		

switch#
！可以看到，已经新增了一个"VLAN100"的信息

第七步：实验结束后，取消 enable 密码。

如果不取消 enable 密码，下一批的同学将没有办法做实验，因此，所有个人在实验中设定的密码都应该在实验完成之后取消，为后面实验的同学带来方便，这也是一个网络工程师基本的素质。

switch(Config)#no enable password level admin
Input password:*****
switch(Config)#

七、注意事项和排错

1. 特定的命令存在于特定的配置模式下。大家在进行配置时不仅仅需要输入正确的命

令，还需要知道该命令是否是在正确的配置模式下。

2. 当用户不知道该命令是否正确时，可以使用"？"来咨询交换机。

八、思考题

1. 为什么 enable 密码在 show 命令显示时，不是出现配置的密码，而是一大堆不认识的字符？

2. 当不能确定一个命令是否存在于某个配置模式下时，应该怎么查询？

九、课后练习

1. 进入各个配置模式并退出。
2. 设置特权用户配置模式的 enable 密码为"digitalchina"。
3. 实验结束后，一定要取消 enable 密码。

十、相关配置命令详解

交换机的配置模式有以下 6 种，如图 1-13 所示。

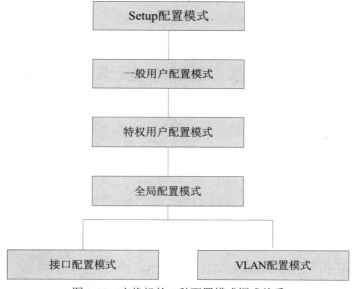

图 1-13　交换机的 6 种配置模式框式关系

1. Setup 配置模式

交换机出厂第一次启动时会自动进入 Setup 配置模式。

Setup 配置大多是以菜单的形式出现的，在 Setup 配置模式中可以做一些交换机最基本的配置，譬如修改交换机提示符、配置交换机 IP 地址和启动 Web 服务等。

更多情况下，为了配置更复杂的网络环境，我们经常直接跳出 Setup 模式，而使用命令行方式进行配置。用户从 Setup 配置模式退出后，进入到 CLI 配置界面。

Setup 模式所做的所有配置在 CLI 配置界面中都可以配置。

并不是所有的交换机都支持 Setup 配置模式。

2. 一般用户配置模式

用户进入 CLI 界面，首先进入的就是一般用户配置模式，提示符为"Switch>"，符号">"为一般用户配置模式的提示符。当用户从特权用户配置模式使用命令 exit 退出时，可以回到一般用户配置模式。

在一般用户配置模式下有很多限制，用户不能对交换机进行任何配置，只能查询交换机的时钟和交换机的版本信息。

所有的交换机都支持一般用户配置模式。

3. 特权用户配置模式

在一般用户配置模式下使用 Enable 命令，如果已经配置了进入特权用户的口令，则输入相应的特权用户口令，即可进入特权用户配置模式"Switch#"。当用户从特权用户配置模式使用 exit 退出时，也可以回到一般用户配置模式。另外交换机提供"Ctrl+Z"的快捷键，使得交换机在任何配置模式(一般用户配置模式除外)，都可以退回到特权用户配置模式。

所有的交换机都支持特权用户配置模式。

4. 全局配置模式

进入特权用户配置模式后，只需使用命令 Config，即可进入全局配置模式"Switch(Config) #"。当用户在其他配置模式，如接口配置模式、VLAN 配置模式时，可以使用命令 exit 退回到全局配置模式。

5. 接口配置模式

在全局配置模式下，使用命令 Interface 就可以进入到相应的接口配置模式。交换机操作系统提供了以下两种端口类型。

(1) CPU 端口：二层交换机中，创建的第一个 vlan 接口的 vlan 被称为管理 vlan，管理 vlan 接口将拥有 CPU 的 MAC 地址。

(2) 以太网端口。

因此就有两种接口的配置模式，如表 1-1 所示。

表 1-1 接口配置模式

接口类型	进入方式	提示符	可执行操作	退出方式
CPU 端口	在全局配置模式下，输入命令 interface vlan 1	Switch(config-vlan)#	配置交换机的 IP 地址，设置管理 vlan	使用 exit 命令即可退回全局配置模式
以太网端口	在全局配置模式下，输入命令 interface ethernet <interface-list>	Switch(config-if)#	配置交换机提供的以太网接口的双工模式、速率、广播抑制等	使用 exit 命令即可退回全局配置模式

6. VLAN 配置模式

在全局配置模式下，使用命令 VLAN<vlan-id>就可以进入到相应的 VLAN 配置模式。在"CS-3926S"中输入所需创建的 VLAN 号，即可进入此 VLAN 的配置模式。在 VLAN 配置模式，用户可以配置本 VLAN 的成员以及各种属性。

config

命令：config [terminal]。

功能：从特权用户配置模式进入到全局配置模式。

参数：[terminal]表示进行终端配置。

命令模式：特权用户配置模式。

举例：

Switch#config

enable

命令：enable。

功能：用户使用 enable 命令，从普通用户配置模式进入特权用户配置模式。

命令模式：一般用户配置模式。

使用指南：为了防止非特权用户的非法访问，在从普通用户配置模式进入到特权用户配置模式时，要进行用户身份验证，即需要输入特权用户口令，输入正确的口令，则进入特权用户配置模式，否则保持普通用户配置模式不变。特权用户口令的设置为全局配置模式下的命令 enable password。

举例：

Switch>enable
password： ***** (admin)
Switch#

相关命令：enable password。

enable password

命令：enable password。

功能：修改从普通用户配置模式进入特权用户配置模式的口令，输入本命令后直接回车将出现<Current password>、<New password>参数，需要用户配置。

参数：<Current password>为原来的密码，最长不超过 16 个字符；<New password>为新的密码，最长不超过 16 个字符；<Confirm new password>为确认新密码，值应与新密码一样，否则需要重新设置密码。

命令模式：全局配置模式。

默认情况：系统默认的特权用户口令为空。当用户是首次配置出现要输入原密码的提示时，直接回车即可。

使用指南：配置特权用户口令，可以防止非特权用户的非法侵入，建议网络管理员在首次配交换机时就设定特权用户口令。另外当管理员需要长时间离开终端屏幕时，最好执行 exit 命令退出特权用户配置模式。

举例：设置特权用户的口令为 admin。

Switch(Config)#enable password
Current password: （首次配置，没有设置任何口令，直接回车）
New password:*****   （设置新的口令为 admin）
Confirm new password:*****(确认新的口令 admin)
Switch(Config)#

相关命令：enable。

exit

命令：exit。

功能：从当前模式退出，进入上一个模式，如在全局配置模式使用本命令退回到特权用户配置模式，在特权用户配置模式使用本命令退回到一般用户配置模式等。

命令模式：各种配置模式。

举例：

Switch#exit
Switch>

配置分级介绍

为了对网络进行有效的保护，允许用户以不同的身份登录交换机进行配置，允许对不同的身份设置不同的密码，不同的身份在配置交换机时有不同的权限。目前 DCN 的交换机可以设置 visitor 和 admin 两种身份。两种登录身份的配置权限区别如表 1-2 所示。

表 1-2　两种登录身份的配置权限

登录身份	配置权限
visitor	绝大部分 show 命令以及 ping、traceroute、clear 等命令。该身份无法进入 config 模式
admin	所有命令

实验三　交换机 CLI 界面调试技巧

一、实验目的

1. 熟悉交换机 CLI 界面。
2. 了解基本的命令格式。

3. 了解部分调试技巧。

二、应用环境

所有其他的实验都需要使用到本实验所讲述的内容，熟悉本实验，将会对其他实验的操作带来方便。

三、实验设备

1. DCS-3926S 交换机　　1 台。
2. PC 机　　1 台。
3. Console 线　　1 根。

四、实验拓扑

该实验拓扑结构如图 1-14 所示。

图 1-14　实验拓扑图

五、实验要求

1. 熟悉帮助功能。
2. 了解交换机对输入的检查。
(1) 成功返回信息。
(2) 错误返回信息。
3. 熟练使用不完全匹配功能。
4. 熟悉以下常用配置技巧。
(1) 命令简写。
(2) 命令完成。
(3) 命令查询。
(4) 否定命令的作用。

(5) 命令历史。

六、实验步骤

第一步："?"的使用。

```
switch#show v?                  ！查看 v 开头的命令
version    vlan                 ！只有两条 show version 和 show vlan
switch#show version             ！查看交换机版本信息
```

第二步：查看错误信息。

```
switch#show v                   ！直接输入 show v，回车
> Ambiguous command!
switch#                         ！根据已有输入可以产生至少两种不同的解释
switch#show valn                ！show vlan 写成了 show valn
> Unrecognized command or illegal parameter!   ！不识别的命令
switch#
```

第三步：不完全匹配。

```
switch#show ver                 ！应该是 show version，没有输全，但是无歧义即可
   DCS-3926S Device, Aug 23 2005 09:35:31
   HardWare version is 1.01
   SoftWare version is DCS-3926S_6.1.12.0
   DCNOS version is DCNOS_5.1.35.42
   BootRom version is DCS-3926S_1.2.0
   Copyright (C) 2001-2005 by Digital China Networks Limited.
   All rights reserved.
   System up time: 0 days, 0 hours, 22 minutes, 43 seconds.
switch#
```

第四步：Tab 的用途。

```
switch#show v                   ！show v 按 Tab 键，出错，因为有 show vlan，有歧义
> Ambiguous command!
switch#show ver                 ！show ver 按 Tab 键补全命令
   DCS-3926S Device, Aug 23 2005 09:35:31
   HardWare version is 1.01
   SoftWare version is DCS-3926S_6.1.12.0
   DCNOS version is DCNOS_5.1.35.42
   BootRom version is DCS-3926S_1.2.0
   Copyright (C) 2001-2005 by Digital China Networks Limited.
   All rights reserved.
   System up time: 0 days, 0 hours, 35 minutes, 56 seconds.
switch#
```

只有当前命令正确的情况下才可以使用 Tab 键，也就是说一旦命令没有输全，但是 Tab 键又没有起作用时，就说明当前的命令中出现了错误，或者命令错误，或者参数错误等，需要仔细排查。

第五步：否定命令"no"。

```
switch#config                                    ！进入全局配置模式
switch(Config)#vlan 10                           ！创建 vlan 10 并进入 vlan 配置模式
switch(Config-Vlan10)#exit                       ！退出 vlan 配置模式
switch(Config)#show vlan                         ！查看 vlan
> Unrecognized command or illegal parameter!     ！该命令不在全局配置模式下
switch(Config)#exit                              ！退出全局配置模式
switch#show vlan                                 ！查看 vlan 信息
```

VLAN	Name	Type	Media	Ports	
1	default	Static	ENET	Ethernet0/0/1	Ethernet0/0/2
				Ethernet0/0/3	Ethernet0/0/4
				Ethernet0/0/5	Ethernet0/0/6
				Ethernet0/0/7	
				Ethernet0/0/9	Ethernet0/0/10
				Ethernet0/0/11	Ethernet0/0/12
				Ethernet0/0/13	Ethernet0/0/14
				Ethernet0/0/15	Ethernet0/0/16
				Ethernet0/0/17	Ethernet0/0/18
				Ethernet0/0/19	Ethernet0/0/20
				Ethernet0/0/21	Ethernet0/0/22
				Ethernet0/0/23	Ethernet0/0/24
10	VLAN0010	Static	ENET		！有 vlan 10 的存在

```
switch#config
switch(Config)#no vlan 10                        ！使用 no 命令删掉 vlan 10
switch(Config)#exit
switch#show vlan
```

VLAN	Name	Type	Media	Ports	
----	------------	----------	--------	--	
1	default	Static	ENET	Ethernet0/0/1	Ethernet0/0/2
				Ethernet0/0/3	Ethernet0/0/4
				Ethernet0/0/5	Ethernet0/0/6
				Ethernet0/0/7	Ethernet0/0/8
				Ethernet0/0/9	Ethernet0/0/10
				Ethernet0/0/11	Ethernet0/0/12
				Ethernet0/0/13	Ethernet0/0/14
				Ethernet0/0/15	Ethernet0/0/16
				Ethernet0/0/17	Ethernet0/0/18
				Ethernet0/0/19	Ethernet0/0/20
				Ethernet0/0/21	Ethernet0/0/22
				Ethernet0/0/23	Ethernet0/0/24

switch#   ! vlan 10 不见了，已经删掉了

交换机中大部分命令的逆命令都是采用 no 命令的模式，还有一种否定的模式是 enable，和 disable 的相反。

第六步：使用上下光标键↑、↓来选择已经输过的命令来节省时间。

七、课后练习

本实验的内容需要课后大量练习 vlan 的划分及端口的验证联通。

八、相关配置命令详解

配置语法

交换机为用户提供了各种各样的配置命令，尽管这些配置命令的形式各不一样，但它们都遵循交换机配置命令的语法。以下是交换机提供的通用命令格式：

cmdtxt <*variable*> {enum1 | enum2 } [option]

语法说明：黑体字 **cmdtxt** 表示命令关键字；<*variable*>表示参数为变量；{enum1 | …|enumN }表示在参数集 enum1~enumN 中必须选一个参数；[option]中的"[ ]"表示该参数为可选项。在各种命令中还会出现"< >""{ }""[ ]"符号的组合使用，如：[<variable>]，{enum1 <variable>| enum2}，[option1 [option2]]等。

下面是几种配置命令语法的具体分析：

- show version，没有任何参数，属于只有关键字没有参数的命令，直接输入命令即可。
- vlan <*vlan-id*>，输入关键字后，还需要输入相应的参数值。
- duplex {auto|full|half}，此类命令用户可以输入 duplex half 或者 duplex full 或者 duplex auto。
- snmp-server community {ro|rw} <string>，出现以下几种输入情况：

snmp-server community ro <string>
snmp-server community rw <string>

支持快捷键

交换机为方便用户的配置，特别提供了多个快捷键，如上、下、左、右键及删除键 Backspace 等。如果超级终端不支持上下光标键的识别，可以使用 Ctrl+P 和 Ctrl+N 来替代，如表 1-3 所示。

表 1-3 常用的快捷键

按 键	功 能
删除键 Backspace	删除光标所在位置的前一个字符，光标前移
上光标键 "↑"	显示上一个输入命令。最多可显示最近输入的 10 个命令
下光标键 "↓"	显示下一个输入命令。当使用上光标键回溯到以前输入的命令时，也可以使下光标键退回到相对于前一个命令的下一个命令
左光标键 "←"	光标向左移动一个位置

(续表)

按键	功能
右光标键 "→"	光标向右移动一个位置
Ctr+P	相当于上光标键 "↑" 的作用
Ctr+N	相当于下光标键 "↓" 的作用
Ctr+Z	从其他配置模式(一般用户配置模式除外)直接退回到特权用户模式
Ctr+C	打断交换机 ping 其他主机的进程
Tab 键	当输入的字符串可以无冲突的表示命令或关键字时，可以使用 Tab 键将其补充成完整的命令或关键字

帮助功能

交换机为用户提供了两种方式获取帮助信息，其中一种方式为使用"help"命令，另一种方式为"?"方式。

help

命令：help

功能：输出有关命令解释器帮助系统的简单描述。

命令模式：各种配置模式。

使用指南：交换机提供随时随地的在线帮助，help 命令则显示关于整个帮助体系的信息，包括完全帮助和部分帮助，用户可以随时输入"?"获取在线帮助。

举例：

```
Switch>help
enable              -- Enable Privileged mode
 exit               -- Exit telnet session
 help               -- help
 show               -- Show running system information
```

两种方式的使用方法和功能如表 1-4 所示。

表 1-4 帮助功能的使用

帮助	使用方法及功能
help	在任一命令模式下，输入"help"命令均可获取有关帮助系统的简单描述
"?"	(1) 在任一命令模式下，输入"?"获取该命令模式下的所有命令及其简单描述 (2) 在命令的关键字后，输入以空格分隔的"?"，若该位置是参数，会输出该参数类型、范围等描述；若该位置是关键字，则列出关键字的集合及其简单描述；若输出"<cr>"，则此命令已输入完整，在该处按 Enter 键即可 (3) 在字符串后紧接着输入"?"，会列出以该字符串开头的所有命令

对输入的检查

(1) 成功返回信息

通过键盘输入的所有命令都要经过 Shell 的语法检查。当用户正确输入相应模式下的命令后，命令执行成功，不会显示信息。

(2) 错误返回信息

常见的错误返回信息如表 1-5 所示。

表 1-5 常见的错误返回信息

输出错误信息	错误原因
Unrecognized command or illegal parameter	命令不存在，或者参数的范围、类型、格式有错误
Ambiguous command	根据已有输入可以产生至少两种不同的解释
Invalid command or parameter	命令解析成功，但没有任何有效的参数记录
Shell Task error…	多任务时，新的 shell 任务启动失败
This command is not exist in current mode	命令可解析，但当前模式下不能配置该命令
Please configurate precursor command "*" at first!	当前输入可以被正确解析，但其前导命令尚未配置
Syntax error: missing ""before the end of command line!	输入中使用了引号，但没有成对出现

支持不完全匹配

绝大部分交换机的 Shell 支持不完全匹配的搜索命令和关键字，当输入无冲突的命令或关键字时，Shell 就会正确解析；有冲突的时候会显示"Ambiguous command"。

例如：对特权用户配置命令"show interface ethernet 1"，只要输入"sh in e 1"即可。

再如：对特权用户配置命令"show running-config"，如果仅输入"sh r"，系统会报"> Ambiguous command!"，因为 Shell 无法区分"show r"是"show rom"命令还是"show running-config"命令，因此必须输入"sh ru"，Shell 才会正确解析。

常用配置技巧

(1) 命令简写

在输入一个命令时可以只输入各个命令字符串的前面部分，只要长到系统能够与其他命令关键字区分即可。例如，如果输入"logging console"命令，可只需输入"logging c"，系统会自动进行识别。如果输入的缩写命令太短，无法与别的命令区分，系统会提示继续输入后面的字符。

(2) 命令完成

如果在输入一个命令字符串的部分字符后输入 Tab 键，系统会自动显示该命令的剩余字符串形成一个完整的命令。例如在输入"log"后输入 Tab 键，系统会自动补成"logging"。当然，所输入的部分字符也需要足够长，以区分不同的命令。

(3) 命令查询

如果知道一个命令的部分字符串，也可以通过在部分字符串后面输入"?"来显示匹配该字符串的所有命令，例如输入"s?"将显示以 s 开头的所有关键字。

```
Console#show s?
Snmp    startup-config system
```

(4) 否定命令的作用

对于许多配置命令，可以输入前缀 no 来取消一个命令的作用或者是将配置重新设置为默认值。例如 logging 命令会将系统信息传送到主机服务器，为了禁止传送，可输入 no logging 命令。本手册将会描述所有可应用命令的否定效果。

(5) 命令历史

交换机可以记忆已经输入的命令，用户可以用"Ctrl+P"调出已经输入的命令，也可以用"show history"来显示已经输入的命令列表。

实验四　交换机恢复出厂设置及其基本配置

一、实验目的

1. 了解交换机的文件管理。
2. 了解什么时候需要将交换机恢复成出厂设置。
3. 了解交换机恢复出厂设置的方法。
4. 了解交换机的一些基本配置命令。

二、应用环境

1. 实际环境如下。

(1) 教学楼的 DCS-3926S 坏了，网络管理员把实验楼的一台交换机拿过去先用着。这台交换机的配置是按照实验楼的环境设置的，我需要改成教学楼的环境，一条条修改比较麻烦，也不能保证正确，不如清空交换机的所有配置，恢复到刚刚出厂的状态。

(2) 我正在配置一台 DCS-3926S，做了很多功能的配置，完成之后发现它不能正常工作。问题出在哪里了？我检查了很多遍都没有发现错误。排错的难度远远大于重新做配置，不如清空交换机的所有配置，恢复到刚刚出厂的状态。

2. 实验环境如下。

上一节网络实验课的同学们刚刚做完实验，已经离去。桌上的交换机他们已经配置过，我通过 show run 命令发现他们对交换机做了很多的配置，有些我能看明白，有些我看不明白。为了不影响我这节课的实验，我必须把他们做的配置都删除，最简单的方法就是清空配置，恢复到刚刚出厂的状态，让交换机的配置成为一张白纸，这样我就能按照自己的思路进行配置，也能更清楚地了解我的配置是否生效，是否正确。

三、实验设备

1. DCS-3926S 交换机　　1 台。
2. PC 机　　1 台。
3. Console 线　　1 根。

四、实验拓扑

该实验拓扑结构如图 1-15 所示。

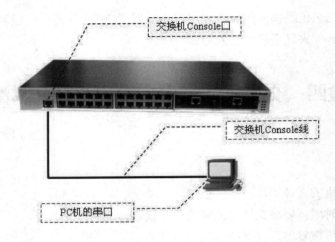

图 1-15 实验拓扑图

五、实验要求

1. 先给交换机设置 enable 密码,确定 enable 密码设置成功。
2. 对交换机做恢复出厂设置,重新启动后发现 enable 密码消失,表明恢复成功。
3. 了解 show flash 命令以及显示内容。
4. 了解 clock set 命令以及显示内容。
5. 了解 hostname 命令以及显示内容。
6. 了解 language 命令以及显示内容。

六、实验步骤

第一步:为交换机设置 enable 密码。(详见实验二)

switch>enable
switch#config t   !进入全局配置模式
switch(Config)#enable password level admin
Current password: !原密码为空,直接回车
New password:*****
Confirm new password:*****
switch(Config)#exit
switch#write
switch#

验证配置:
验证方法 1:重新进入交换机

```
switch#exit                                    ! 退出特权用户配置模式
switch>
switch>enable                                  ! 进入特权用户配置模式
Password:*****
switch#
```

验证方法 2：show 命令来查看

```
switch#show running-config
Current configuration:
!
     enable password level admin 827ccb0eea8a706c4c34a16891f84e7b   ! 该行显示了已经为交换机配置了 enable 密码。
     hostname switch
!
!
Vlan 1
     vlan 1
!
!
……                                             ! 省略部分显示
```

第二步：清空交换机的配置。

```
switch>enable                                  ! 进入特权用户配置模式
switch#set default                             ! 使用 set default 命令
Are you sure? [Y/N] = y                        ! 是否确认？
switch#write                                   ! 清空 startup-config 文件
switch #show startup-config                    ! 显示当前的 startup-config 文件
This is first time start up system.            ! 系统提示此启动文件为出厂默认配置
switch#reload                                  ! 重新启动交换机
Process with reboot? [Y/N] y
```

验证测试：

验证方法 1：重新进入交换机

```
switch>
switch>enable
switch#                                        ! 已经不需要输入密码就可进入特权模式
```

验证方法 2：show 命令来查看

```
switch#show running-config
Current configuration:
!
     hostname switch                           !  已经没有 enable 密码显示了
```

!
Vlan 1
　　vlan 1
!
……　　　　　　　　　　　　　　　　　　　! 省略部分显示

第三步：show flash 命令。

```
switch#show flash
file name              file length
nos.img                1720035 bytes        ! 交换机软件系统
startup-config         0 bytes              ! 启动配置文件
running-config         783 bytes            ! 当前配置文件
switch#
```

第四步：设置交换机系统日期和时钟。

```
switch#clock set ?                                ! 使用？查询命令格式
  <HH:MM:SS>                     -- Time
switch#clock set 15:29:50                         ! 配置当前时间
Current time is MON JAN 01 15:29:50 2001          ! 配置完即有显示，注意年份不对
switch#clock set 15:29:50 ?                       ! 使用？查询，原来命令没有结束
  <YYYY.MM.DD>                   -- Date <year:2000-2035>
  <CR>
switch#clock set 15:29:50 2006.01.16              ! 配置当前年月日
Current time is MON JAN 16 15:29:50 2006          ! 正确显示
```

验证配置：

```
switch#show clock                                 ! 再用 show 命令验证
Current time is MON JAN 16 15:29:55 2006
switch#
```

第五步：设置交换机命令行界面的提示符(设置交换机的姓名)。

```
switch#
switch#config
switch(Config)#hostname DCS-3926S-BD1             ! 配置姓名
DCS-3926S-BD1(Config)#exit                        ! 无须验证，即配即生效
DCS-3926S-BD1#
DCS-3926S-BD1#
```

第六步：配置显示的帮助信息的语言类型。

```
DCS-3926S-BD1#language ?
  chinese                        -- Chinese
  english                        -- English
```

```
DCS-3926S-BD1#language chinese
DCS-3926S-BD1#language ?              ！请注意再使用?时，帮助信息已经成了中文。
    chinese                -- 汉语
    english                -- 英语
```

七、注意事项和排错

1. 恢复出厂设置 set default 后一定要 write，重新启动后生效。
2. 这几个命令中，hostname 命令是在全局配置模式下配置的。

八、配置序列

```
DCS-3926S-BD1#show running-config
Current configuration:
!
    hostname DCS-3926S-BD1    ！上述的配置只有 hostname 命令在 show run 中可以显示
!
!
Vlan 1
    vlan 1
!
Interface Ethernet0/0/1
……                                   ！省略部分
Interface Ethernet0/0/23
!
Interface Ethernet0/0/24
!
DCS-3926S-BD1#
```

九、思考题

1. 为什么第三步中 show flash 的显示中，startup-config 文件的大小是 0 bytes？
2. 怎样才能将 startup-config 文件和 running-config 文件保持一致？

十、课后练习

1. 请为交换机设置 enable 密码为 digitalchina。
2. 请把交换机的时钟设置为当前时间。
3. 请为交换机设置名称为 digitalchina-3926S。
4. 请把交换机的帮助信息设置为中文。
5. 请把交换机恢复到出厂设置。

十一、相关配置命令详解

clock set

命令：clock set <HH:MM:SS> <YYYY/MM/DD>。

功能：设置系统日期和时钟。

参数：<HH:MM:SS>为当前时钟，HH 取值范围为 0~23，MM 和 SS 取值范围为 0~59；YYYY/MM/DD>为当前年、月和日，YYYY 取值范围为 2000~2100，MM 取值范围为 1~12，DD 取值范围为 1~31。

命令模式：特权用户配置模式。

默认情况：系统启动时默认为 2001 年 1 月 1 日 0：0：0。

使用指南：交换机在断电后不能继续计时，因此在要求使用确切时间的应用环境中，必须先设定交换机当前的日期和时间。

举例：设置交换机当前日期为 2002年8月1日23时0分0秒。

Switch#clock set 23:0:0 2002.8.1

相关命令：show clock。

hostname

命令：hostname <hostname>。

功能：设置交换机命令行界面的提示符。

参数：<hostname>为提示符的字符串。

命令模式：全局配置模式。

默认情况：系统默认提示符为"DCS-3926S"。

使用指南：通过本命令用户可以根据实际情况设置交换机命令行的提示符。

举例：设置提示符为Test。

Switch(Config)#hostname Test
Test(config)#

language

命令：language {chinese|english}。

功能：设置显示的帮助信息的语言类型。

参数：chinese为中文显示；english为英文显示。

命令模式：特权用户配置模式。

默认情况：系统默认是英文显示。

使用指南：DCS-3926S提供了两种语言的帮助信息，用户可根据自己的喜好选择语言类型。系统若重启后，帮助显示信息恢复为英文显示。

reload

命令：reload。

功能：热启动交换机。

命令模式：特权用户配置模式。

使用指南：用户可以通过本命令，在不关闭电源的情况下，重新启动交换机。

set default

命令：set default。

功能：恢复交换机的出厂设置。

命令模式：特权用户配置模式。

使用指南：恢复交换机的出厂设置，即用户对交换机做的所有配置都消失，用户重新启动交换机后，出现的提示与交换机首次上电一样。

注意：

配置本命令后，必须执行write命令，进行配置保留后重启交换机即可使交换机恢复到出厂设置。

举例：

Switch#set default
Are you sure? [Y/N] = y
Switch#write
Switch#reload

show flash

命令：show flash。

功能：显示保存在 flash 中的文件及大小。

命令模式：特权用户配置模式。

举例：查看 flash 中文件及大小。

```
Switch#show flash
file name              file length
nos.img                1122380 bytes
startup-config         1061 bytes
running-config         1061 bytes
Switch#
```

show running-config

命令：show running-config。

功能：显示当前运行状态下生效的交换机参数配置。

默认情况：对于正在生效的配置参数，如果与默认工作参数相同，则不显示。

命令模式：特权用户配置模式。

使用指南：当用户完成一组配置后，需要验证是否配置正确，则可以执行 show running-config令来查看当前生效的参数。

举例：

Switch#show running-config

show startup-config

命令：show startup-config。

功能：显示当前运行状态下写在 Flash Memory 中的交换机参数配置，通常也是交换机下次上电启动时所用的配置文件。

默认情况：从Flash中读出的配置参数，如果与默认工作参数相同，则不显示。

命令模式：特权用户配置模式。

使用指南：show running-config和show startup-config命令的区别在于，当用户完成一组配置之后，通过show running-config可以看到配置增加了，而通过show startup-config却看不出配置的变化。但若用户通过write命令，将当前生效的配置保存到Flash Memory中时，show running-config的显示与show startup-config的显示结果一致。

实验五　使用 Telnet 方式管理交换机

一、实验目的

1. 了解什么是带内管理。
2. 熟练掌握如何使用 Telnet 方式管理交换机。

二、应用环境

学校有 20 台交换机支撑着校园网的运营，这 20 台交换机分别放置在学校的不同位置。作为网络管理员需要对这 20 台交换机做管理，通过前面学习的知识，我们可以通过带外管理的方式也就是通过 Console 口去管理，那么管理员需要捧着自己的笔记本电脑，并且带着 Console 线去学校的不同位置去调试每台交换机，十分麻烦。

校园网既然是互联互通的，在网络的任何一个信息点都应该能访问其他的信息点，我们为什么不通过网络方式来调试交换机呢？通过 Telnet 方式，管理员就可以坐在办公室中方便地调试全校所有的交换机。

Telnet 方式和下个实验中的 Web 方式都是交换机的带内管理方式。

提供带内管理方式可以使连接在交换机中的某些设备具备管理交换机的功能。当交换机的配置出现变更，导致带内管理失效时，必须使用带外管理对交换机进行配置管理。

三、实验设备

1. DCS-3926S交换机　　1台。
2. PC机　　1台。
3. Console线　　1根。
4. 直通网线　　1根。

四、实验拓扑

该实验拓扑结构如图1-16所示。

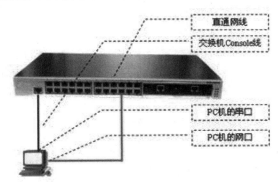

图1-16　实验拓扑图

五、实验要求

1. 按照拓扑图连接网络。
2. PC和交换机的24口用网线相连。
3. 交换机的管理IP为192.168.2.100/24。
4. PC网卡的IP地址为192.168.2.101/24。

六、实验步骤

第一步：交换机恢复出厂设置，设置正确的时钟和标识符(详见实验四)。

switch#set default
Are you sure? [Y/N] = y
switch#write
switch#reload
Process with reboot? [Y/N] y
switch#clock set 15:29:50 2006.01.16
Current time is MON JAN 16 15:29:50 2006
switch#
switch#config
switch(Config)#hostname DCS-3926S
DCS-3926S(Config)#exit
DCS-3926S#

第二步：给交换机设置 IP 地址即管理 IP。

DCS-3926S#config
DCS-3926S(Config)#interface vlan 1   ！进入 vlan 1 接口
02:20:17: %LINK-5-CHANGED: Interface Vlan1, changed state to UP
DCS-3926S(Config-If-Vlan1)#ip address 192.168.2.100 255.255.255.0 ！配置地址
DCS-3926S(Config-If-Vlan1)#no shutdown   ！激活 vlan 接口
DCS-3926S(Config-If-Vlan1)#exit
DCS-3926S(Config)#exit
DCS-3926S#

验证配置：

DCS-3926S#show run
Current configuration:
!
    hostname DCS-3926S
!
Vlan 1
    vlan 1
!
Interface Ethernet0/0/1
……
Interface Ethernet0/0/24
!
interface Vlan1
    interface vlan 1
    ip address 192.168.2.100 255.255.255.0   ！已经配置好交换机 IP 地址
!
DCS-3926S#

第三步：为交换机设置授权 Telnet 用户。

DCS-3926S#config
DCS-3926S(Config)#telnet-user xuxp password 0 digital
DCS-3926S(Config)#exit
DCS-3926S#

验证配置：

DCS-3926S#show run
Current configuration:
!
    hostname DCS-3926S
!
    telnet-user xuxp password 0 digital

!
Vlan 1
    vlan 1
!
Interface Ethernet0/0/1
……
Interface Ethernet0/0/24
!
interface Vlan1
    interface vlan 1
    ip address 192.168.2.100 255.255.255.0
!
DCS-3926S#

第四步：配置主机的 IP 地址，在本实验中要与交换机的 IP 地址在一个网段。
主机 IP 配置如图 1-17 所示。

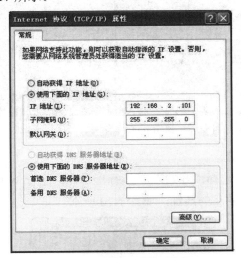

图 1-17　IP 地址配置

验证配置：可以在PC主机的DOS命令行中使用ipconfig命令查看IP地址配置，结果如图1-18所示。

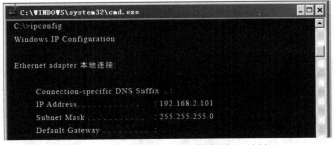

图 1-18　使用 ipconfig 命令查看 IP 地址

第五步：验证主机与交换机是否连通。

验证方法 1：在交换机中 ping 主机。

DCS-3926S#ping 192.168.2.101
Type ^c to abort.
Sending 5 56-byte ICMP Echos to 192.168.2.101, timeout is 2 seconds.
!!!!!
Success rate is 100 percent (5/5), round-trip min/avg/max = 1/1/1 ms
DCS-3926S#

很快出现 5 个 "!" 表示已经连通。

验证方法 2：在主机 DOS 命令行中 ping 交换机，若出现如图 1-19 所示的结果则表示连通。

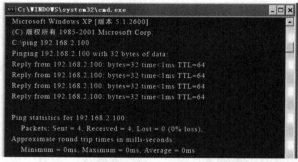

图 1-19　使用 ping 命令查看网络连通性

第六步：使用 Telnet 登录。

打开 Windows 系统，选择"开始"|"运行"命令，运行 Windows 自带的 Telnet 客户端程序，并且指定 Telnet 的目的地址，如图 1-20 所示。

图 1-20　执行 telnet 命令

需要输入正确的登录名和口令，登录名是 xuxp，口令是 digital。Telnet 登录界面如图 1-21 所示。

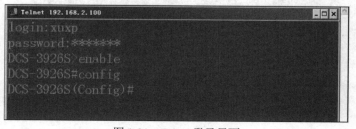

图 1-21　Telnet 登录界面

可以对交换机做进一步配置,本实验完成。

七、注意事项和排错

1. 默认情况下,交换机所有端口都属于 vlan1,因此我们通常把 vlan1 作为交换机的管理 vlan,因此 vlan1 接口的 IP 地址就是交换机的管理地址。

2. 密码只能是 1~8 个字符。

3. 删除一个 telnet 用户可以在 config 模式下使用 no telnet-user 命令。

八、配置序列

```
DCS-3926S#show running-config
Current configuration:
!
    hostname DCS-3926S
!
    telnet-user xuxp password 0 digital
!
Vlan 1
    vlan 1          !
!
Interface Ethernet0/0/1
......
Interface Ethernet0/0/24 !
interface Vlan1
    interface vlan 1
    ip address 192.168.2.100 255.255.255.0
!
DCS-3926S#
```

九、思考题

1. 二层交换机的 IP 地址可以配置多少个,为什么?

2. 能不能为 vlan2 配置 IP 地址?

3. telnet-user xuxp password 0 digital 中把"0"换成"7"会是什么现象?

十、课后练习

1. 删除 xuxp 用户(不准用 set default)。

2. 设置交换机的管理 IP 为 10.1.1.1 255.255.255.0。

3. 使用用户名 aaa,密码 bbb,并且选择"7"作为参数配置 Telnet 功能。

十一、相关配置命令详解

三层接口介绍：在 DCS-3926 交换机上只能创建一个管理三层接口。三层接口并不是实际的物理接口，它是一个虚拟的接口。三层接口是在 VLAN 的基础上创建的。三层接口可以包含一个或多个二层接口(它们同属于一个 VLAN)，但也可以不包含任何二层接口。三层接口包含的二层接口中，需要至少有一个是 UP 状态，三层接口才是 UP 状态，否则为 DOWN 状态。每一个三层接口有自己独立的 MAC 地址，此地址是在三层接口创建时从交换机保留的 MAC 地址中选取的。三层接口是三层协议的基础，在三层接口上可以配置 IP 地址，交换机可以通过配置在三层接口上的 IP 地址，与其他设备进行 IP 协议的传输。

ip address

命令：ip address <ip-address> <mask> [secondary]。
　　　no ip address [<ip-address> <mask>] [secondary]。
功能：设置交换机的 IP 地址及掩码；本命令的 no 操作为删除该 IP 地址配置。
参数：<ip-address>为 IP 地址，点分十进制格式；<mask>为子网掩码，点分十进制格式；[secondary]为表示配置的 IP 地址为从 IP 地址。
默认情况：出厂时交换机无 IP 地址。
命令模式：VLAN 接口配置模式。
使用指南：用户若要为交换机配置 IP 地址，必须首先创建一个 VLAN 接口。
举例：设置 VLAN1 接口的 IP 地址为 10.1.128.1/24。

Switch(Config)#interface vlan 1
Switch(Config-If-Vlan1)#ip address 10.1.128.1 255.255.255.0
Switch(Config-If-Vlan1)#no shut
Switch(Config-If-Vlan1)#exit
Switch(Config)#

相关命令：ip bootp-client enable、ip dhcp-client enable。

interface vlan

命令：interface vlan <vlan-id>
　　　no interface vlan <vlan-id>
功能：创建一个 VLAN 接口，即创建一个交换机的三层接口；本命令的 no 操作为删除交换机的三层接口。
参数：<vlan-id>是已建立的 VLAN 的 VLAN ID。
默认情况：出厂时没有三层接口。
命令模式：全局配置模式。
使用指南：在创建 VLAN 接口(三层接口)前，需要先配置 VLAN，详细内容参见 VLAN 的章节。使用本命令在创建 VLAN 接口(三层接口)的同时，进入 VLAN(三层接口)配置模式。在 VLAN 接口(三层接口)创建好之后，仍然可以使用 interface vlan 命令进入三层接口模式。

举例：在 VLAN 1 上创建一个 VLAN 接口(三层接口)。

Switch (Config)#interface vlan 1

ping

命令：ping [<ip-addr>]

功能：交换机向远端设备发ICMP请求包，检测交换机与远端设备之间是否可达。

参数：<ip-addr>为要ping的目的主机的IP地址，点分十进制格式。

默认情况：发5个ICMP请求包；包大小为56 bytes；超时时间为2秒。

命令模式：特权用户配置模式。

使用指南：当用户输入ping命令后，直接回车，系统提供给用户一种交互式的配置方式，用户可以自定义ping的各项参数值。

举例：

例1：使用 ping 程序的默认参数。

Switch#ping 10.1.128.160
Type ^c to abort.
Sending 5 56-byte ICMP Echos to 10.1.128.160, timeout is 2 seconds.
...!!
Success rate is 40 percent (2/5), round-trip min/avg/max = 0/0/0 ms

上面的例子表示，交换机 ping 某一 IP 地址为 10.1.128.160 的设备，前 3 个 ICMP 请求包在默认超时时间 2 秒内没有收到相应的 ICMP 回应包，即没有 ping 通，而后两个包 ping 通了，成功率为40%。交换机用"."表示 ping 失败，链路不可达；用"!"表示 ping 成功，链路可达。

例 2：使用 ping 程序提供的手段修改 ping 参数，如表 1-6 所示。

Switch#ping ↵
protocol [IP]: ↵
Target IP address：10.1.128.160↵
Repeat count [5]: 100 ↵
Datagram size in byte [56]: 1000 ↵
Timeout in milli-seconds [2000]: 500 ↵
Extended commands [n]: n

表 1-6 Ping 命令参数

显 示 信 息	解　　释
protocol [IP]:	选择 IP 协议的 Ping
Target IP address:	目标设备的 IP 地址
Repeat count [5]	发包的数目，默认为 5
Datagram size in byte [56]	ICMP 包的大小，默认为 56
Timeout in milli-seconds [2000]:	超时时间，单位为毫秒，默认为 2 秒
Extended commands [n]:	是否改变需要其他的选项

shutdown

命令：shutdown。

no shutdown。

功能：关闭指定的以太网端口；本命令的 no 操作为打开端口。

命令模式：端口配置模式。

默认情况：以太网端口默认为打开。

使用指南：当关闭以太网端口时，以太网端口将不发送数据帧，并且在show interface 时显示端口状态为down。

举例：打开0/0/1-8号端口。

Switch(Config)#interface ethernet 0/0/1-8
Switch(Config-Port-Range)#no shutdown

telnet

命令：telnet [<ip-addr>] [<port>]。

功能：以Telnet方式登录到IP地址为<ip-addr>的远程主机。

参数：<ip-addr>为远端主机的IP地址，点分十进制格式；<port>为端口号，取值范围是0~65535。

命令模式：特权用户配置模式。

使用指南：本命令是交换机作为Telnet客户端时使用的，用户通过本命令登录远程主机进行配置。当交换机作为Telnet客户端时，只能与一个远程主机建立TCP连接，如果想与另一个远程主机建立连接，则必须先断开与上一个远程主机的TCP连接。断开与远程主机的连接可以使用快捷键"CTRL+|"。直接输入关键字Telnet后面不加任何参数，用户将进入Telnet配置模式。

举例：交换机Telnet到IP地址为20.1.1.1的远程路由器DCR。

```
Switch#telnet 20.1.1.1 23
Trying 20.1.1.1...
Service port is 23
Connected to 20.1.1.1
login:123
password:***
DCR>
```

telnet-server enable

命令：telnet-server enable。

no telnet-server enable。

功能：打开交换机的Telnet服务器功能；本命令的no操作为关闭交换机的Telnet服务器功能。

默认情况：系统默认打开Telnet服务器功能。

命令模式：全局配置模式。

使用指南：该命令只能在Console下使用，管理员使用本命令允许或拒绝Telnet客户端登录到交换机。

举例：关闭交换机的Telnet服务器功能。

Switch(Config)#no telnet-server enable

telnet server securityip

命令：telnet-server securityip <ip-addr>。

　　　no telnet-server securityip <ip-addr>。

功能：配置交换机作为 Telnet 服务器允许登录的 Telnet 客户端的安全 IP 地址；本命令的 no 操作为删除指定的 Telnet 客户端的安全 IP 地址。

参数：<ip-addr>可以访问本交换机的安全 IP 地址，点分十进制格式。

默认情况：系统默认不配置任何安全 IP 地址。

命令模式：全局模式。

使用指南：没有配置安全 IP 地址前，不限制登录交换机的 Telnet 客户端的 IP 地址；配置安全 IP 地址后，只有安全 IP 地址的主机才能够 Telnet 到交换机进行配置。交换机允许配置多个安全 IP 地址。

举例：设置192.168.2.21为安全IP地址。

Switch(Config)#telnet-server securityip 192.168.2.21

telnet-user

命令：telnet-user <username> password {0|7} <password>。

　　　no telnet-user <username>。

功能：设置Telnet客户端的用户名及口令；本命令的no操作为删除该Telnet用户。

参数：<username>为Telnet客户端用户名，最长不超过16个字符；<password>为登录口令，最长不超过8个字符；0|7分别表示口令不加密显示和加密显示。

命令模式：全局配置模式。

默认情况：系统默认没有设置Telnet客户端的用户名及口令。

使用指南：本命令是交换机作为Telnet服务器时使用的，用户通过本命令设置授权的Telnet客户端。若没有设置授权的Telnet客户端，任何Telnet客户端都不能通过Telnet配置交换机。交换机作为Telnet服务器时，最多允许同时与5个Telnet客户端建立TCP连接。

举例：设置一个名为Antony的Telnet客户端用户，密码为switch。

Switch(Config)#telnet-user Antony password 0 switch

实验六　使用 Web 方式管理交换机

一、实验目的

1. 熟练掌握如何为交换机设置Web方式管理。
2. 熟练掌握如何进入交换机Web管理方式。
3. 了解交换机Web配置界面，并能进行部分操作。

二、应用环境

Web方式，也叫作http方式，和telnet方式一样都可以使管理员做到坐在办公室中不动地方地调试全校所有的交换机。

Web方式比较简单，如果用户不习惯CLI界面的调试，就可以采用Web方式调试。主流的调试界面还是CLI界面，推荐大家要着重学习CLI界面。

三、实验设备

1. DCS-3926S交换机　　1台。
2. PC机　　1台。
3. Console线　　1根。
4. 直通网线　　1根。

四、实验拓扑

该实验拓扑结构如图 1-22 所示。

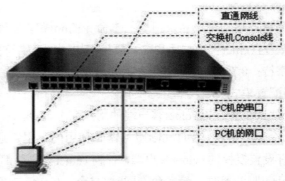

图 1-22　实验拓扑图

五、实验要求

1. 按照拓扑图连接网络。
2. PC和交换机的24口用网线相连。

3. 交换机的管理 IP 为 192.168.2.100/24。
4. PC 网卡的 IP 地址为 192.168.2.101/24。

六、实验步骤

第一步：交换机恢复出厂设置，设置正确的时钟和标识符(详见实验四)。
第二步：给交换机配置管理 IP(详见实验五)。

DCS-3926S#config
DCS-3926S(Config)#interface vlan 1
DCS-3926S(Config-If-Vlan1)#ip address 192.168.2.100 255.255.255.0
DCS-3926S(Config-If-Vlan1)#no shutdown
DCS-3926S(Config-If-Vlan1)#exit
DCS-3926S(Config)#

第三步：启动交换机 Web 服务。

DCS-3926S#config
DCS-3926S(Config)#ip http server
web server is on   ！表明已经成功启动
DCS-3926S(Config)#

第四步：设置交换机授权 HTTP 用户。

DCS-3926S(Config)#web-user admin password 0 digital
DCS-3926S(Config)#

第五步：配置主机的 IP 地址，在本实验中要与交换机的 IP 地址在一个网段(详见实验五)。

配置主机的地址为：192.168.2.101。

第六步：验证主机与交换机是否连通(详见实验五)。
验证方法 1：在交换机中 ping 主机。
验证方法 2：在主机 DOS 命令行中 ping 交换机。
第七步：使用 http 登录。

打开 Windows 系统，选择"开始"|"运行"命令，在"运行"对话框中输入目标地址，如图 1-23 所示。

图 1-23　在 Windows 中打开运行界面

需要输入正确的登录名和口令，登录名是 admin，口令是 digital。登录界面如图 1-24 所示。

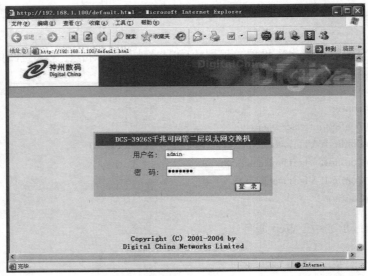

图 1-24　登录界面

下面是交换机 Web 调试界面的主界面，如图 1-25 所示。

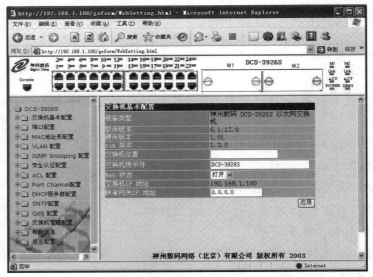

图 1-25　交换机 Web 调试界面的主页面

可以对交换机做进一步配置，本实验完成。

七、注意事项和排错

1. 使用 Telnet 和 Web 方式调试有两个相同的前提条件。
(1) 交换机开启该功能并设置用户。
(2) 交换机和主机之间要互连互通能 ping 通。

2. 有时候交换机的地址配置正确，主机配置也正确，但是就是ping不通。排除硬件问题之后可能的原因是主机的Windows操作系统有防火墙，关闭防火墙即可。

八、配置序列

```
DCS-3926S#show run
Current configuration:
!
    hostname DCS-3926S
!
    telnet-user xuxp password 0 digital
!
Vlan 1
    vlan 1
!
Interface Ethernet0/0/1
......
Interface Ethernet0/0/24
!
interface Vlan1
    interface vlan 1
    ip address 192.168.2.100 255.255.255.0
!
    ip http server
    web-user admin password 0 digital
!
DCS-3926S#
```

九、思考题

1. 如何关闭 Web 服务？
2. 如何删除 Web 用户？

十、课后练习

1. 重新配置一个 Web 用户。
2. 在 Web 界面下，找出前几个实验中配置命令的所在位置，并修改配置。

十一、相关配置命令详解

ip http server

命令：ip http server。
　　　　no ip http server。
功能：使能 Web 配置；本命令的 no 操作为关闭 Web 配置。

命令模式：全局配置模式。

使用指南：Web 配置是给用户提供一个以 HTTP 方式配置的界面。Web 配置的优点是配置直观、形象，容易理解。本命令的作用相当于在 Setup 配置模式的主菜单中选择[2]，进行 Web Server 的配置。

举例：打开 Web Server 功能，使能 Web 配置。

Switch(Config)#ip http server

相关命令：web-user

web-user

命令：web-user <username> password {0|7} <password>。
　　　no web-user <username>。

功能：设置 Web 客户端的用户名及口令；本命令的 no 操作为删除该 Web 客户。

参数：<username>为 Web 访问的授权用户名，最长不超过 16 个字符；<password>为登录口令，最长不超过 8 个字符；0|7 分别表示口令不加密显示和加密显示。

命令模式：全局配置模式。

使用指南：DCS-3926S提供了HTTP方式管理交换机。为了防止非授权用户的Web访问，管理员可以使用本命令配置Web访问的授权用户及口令。

举例：设置一个名为 Admin 的 Web 访问用户，密码为 switch。

Switch(Config)#web-user Admin password 0 switch

相关命令：ip http server

实训一　Telnet 和 Web 方式管理交换机

一、实验目的

1. 熟练掌握如何使用 Telnet 方式管理交换机。
2. 熟练掌握如何为交换机设置 Web 方式管理。
3. 熟练掌握如何进入交换机 Web 管理方式。
4. 了解交换机 Web 配置界面，并能进行部分操作。

二、实验设备

1. DCS-3926S 交换机　　1 台。
2. PC 机　　1 台。
3. 直通网线　　1 根。

三、实验拓扑

该实验拓扑结构如图 1-26 所示。

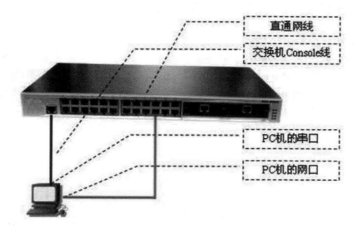

图 1-26 实验拓扑图

四、实验内容

1. 按照拓扑图连接网络。
2. PC 和交换机的 24 口用网线相连。
3. 交换机的管理 IP 为 192.168.2.100/255.255.255.0。
4. PC 网卡的 IP 地址为 192.168.2.101/24。

在本实验室设置 IP 要注意以下 3 点：

1. 本地连接的所有配置均不能修改(除非课堂强调)。
2. 能够修改的只能是本地连接 2 的设置。
3. 测试中使用的 IP 地址段只能使用 2~254 段的地址。

例如：192.168.2.* 255.255.255.0。

不能使用 192.168.1.* 255.255.255.0。

五、实验步骤

第一步：交换机恢复出厂设置，设置正确的时钟和标识符。

switch#set default
Are you sure? [Y/N] = y
switch#write
switch#reload
Process with reboot? [Y/N] y
switch#clock set 15:29:50 2009.*.*
switch#la ch
switch#config
switch(Config)#hostname *****

```
*****(Config)#exit
*****#
```

第二步：给交换机设置 IP 地址即管理 IP。

```
DCS-3926S#config
DCS-3926S(Config)#interface vlan 1                              ！进入 vlan 1 接口
DCS-3926S(Config-If-Vlan1)#ip address 192.168.2.100 255.255.255.0  ！配置地址
DCS-3926S(Config-If-Vlan1)#no shutdown                          ！激活 vlan 接口
DCS-3926S(Config-If-Vlan1)#exit
DCS-3926S(Config)#exit
DCS-3926S#
```

验证配置：

```
DCS-3926S#show run
Current configuration:
!
    hostname DCS-3926S
!
Vlan 1
    vlan 1
!
Interface Ethernet0/0/1
……
Interface Ethernet0/0/24
!
interface Vlan1
    interface vlan 1
    ip address 192.168.2.100 255.255.255.0      ！已经配置好交换机 IP 地址
!
DCS-3926S#
```

第三步：启动交换机 Web 服务。

```
DCS-3926S#config
DCS-3926S(Config)#ip http server
web server is on                    ！表明已经成功启动
DCS-3926S(Config)#
```

第四步：为交换机设置授权 Telnet 和 Web 用户(HTTP 用户)。

```
DCS-3926S#config
DCS-3926S(Config)#telnet-user wl password 0 123
DCS-3926S(Config)#web-user wl2 password 0 1234
DCS-3926S(Config)#
```

第五步：如图 1-27 所示，配置主机的 IP 地址，在本实验中要与交换机的 IP 地址在一个网段。

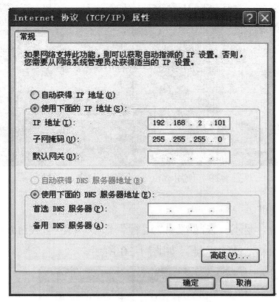

图 1-27　配置 IP 地址

验证配置：

在 PC 主机的 DOS 命令行中使用 ipconfig 命令查看 IP 地址配置，如图 1-28 所示。

图 1-28　使用 ipconfig 查看 IP 地址

第六步：验证主机与交换机是否连通。

验证方法 1：在交换机中 ping 主机。

DCS-3926S#ping 192.168.2.101
Sending 5 56-byte ICMP Echos to 192.168.2.101, timeout is 2 seconds.
!!!!!
Success rate is 100 percent (5/5), round-trip min/avg/max = 1/1/1 ms
DCS-3926S#

很快出现 5 个"!"表示已经连通。

验证方法 2：如图 1-29 所示，在主机 DOS 命令行中 ping 交换机，出现以下显示表示连通。

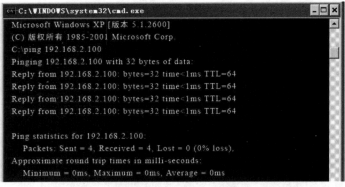

图 1-29 使用 ping 命令测试网络连通性

第七步：使用 Telnet 登录。

打开 Windows 系统，选择"开始"|"运行"命令，运行 Windows 自带的 Telnet 客户端程序，并且指定 Telnet 的目的地址，如图 1-30 所示。

图 1-30 使用 Telnet 命令

需要输入正确的登录名和口令，登录名是 xuxp，口令是 digital。Telnet 登录界面如图 1-31 所示。

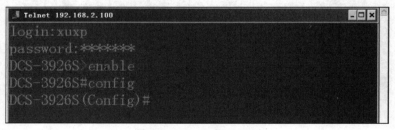

图 1-31 Telnet 登录界面

可以对交换机做进一步配置，本实验完成。

第八步：使用 Web 方式管理交换机。

使用 http://192.168.2.100 登录。

打开 Windows 系统，选择"开始"|"运行"命令，在"运行"对话框里输入目标地址 http://192.168.2.100。

六、注意事项和排错

1. 默认情况下，交换机所有端口都属于 vlan1，因此我们通常把 vlan1 作为交换机的管理 vlan，因此 vlan1 接口的 IP 地址就是交换机的管理地址。
2. 密码只能是 1~8 个字符。
3. 删除一个 telnet 用户可以在 config 模式下使用 no telnet-user 命令。
4. 使用 Telnet 和 Web 方式调试有两个相同的前提条件。
(1) 交换机开启该功能并设置用户。
(2) 交换机和主机之间要互连互通能 ping 通。

七、课后练习

详细步骤概要说明：
1. 选好对应设备，按照拓扑图连接网络。
2. PC 和交换机的 24 口用网线相连。
3. 设置：交换机的管理 IP 为 192.168.2.100/255.255.255.0。
4. 设置：PC 网卡的 IP 地址为 192.168.2.101/24。
5. 设备连通测试：ping 192.168.2.100 –t。
6. 配置授权：
(1) Telnet 登录用户名：wl，密码：123。
(2) 开启 web 的 HTTP 服务：ip http server。
(3) web 登录用户名：wl2，密码：1234。
7. 两种方式分别登录验证。
8. 成功登录后写实验报告。

实验七　交换机 VLAN 划分实验

一、实验目的

1. 了解 VLAN 原理。
2. 熟练掌握二层交换机 VLAN 的划分方法。
3. 了解如何验证 VLAN 的划分。

二、应用环境

学校实验楼中有两个实验室位于同一楼层，一个是计算机软件实验室，一个是多媒体实验室，两个实验室的信息端口都连接在一台交换机上。学校已经为实验楼分配了固定的 IP 地址段，为了保证两个实验室的相对独立，就需要划分对应的 VLAN，使交换机某些端口属于软件实验室，某些端口属于多媒体实验室，这样就能保证它们之间的数据互不干扰，也不影响各自的通信效率。

三、实验设备

1. DCS-3926S 交换机   1 台。
2. PC 机   2 台。
3. Console 线   1 根。
4. 直通网线   2 根。

四、实验拓扑

该实验拓扑结构如图 1-32 所示。

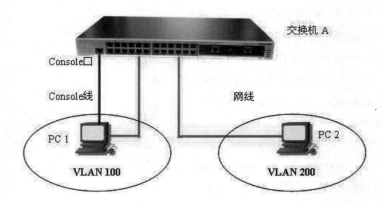

图 1-32　实验拓扑图

使用一台交换机和两台 PC 机，将其中 PC1 作为控制台终端，使用 Console 口配置方式；使用两根网线分别将 PC1 和 PC2 连接到交换机的 RJ-45 接口上。

五、实验要求

在交换机上划分两个基于端口的 VLAN：VLAN100、VLAN200。如表 1-7 所示。

表 1-7　VLAN 端口划分

VLAN	端口成员
100	1~8
200	9~16

使得 VLAN100 的成员能够互相访问，VLAN200 的成员能够互相访问；VLAN100 和 VLAN200 成员之间不能互相访问。

PC1 和 PC2 的网络设置如表 1-8 所示。

表 1-8　设备 IP 网络设置

设备	IP 地址	Mask
交换机 A	192.168.2.11	255.255.255.0
PC1	192.168.2.101	255.255.255.0
PC2	192.168.2.102	255.255.255.0

PC1、PC2 接在 VLAN100 的成员端口 1~8 上,两台 PC 互相可以 ping 通;PC1、PC2 接在 VLAN 的成员端口 9~16 上,两台 PC 互相可以 ping 通;PC1 接在 VLAN100 的成员端口 1~8 上,PC2 接在 VLAN200 的成员端口 9~16 上,则互相 ping 不通。

若实验结果和理论相符,则本实验完成。

六、实验步骤

第一步:交换机恢复出厂设置。

switch#set default
switch#write
switch#reload

第二步:给交换机设置 IP 地址即管理 IP。

switch#config
switch(Config)#interface vlan 1
switch(Config-If-Vlan1)#ip address 192.168.2.11 255.255.255.0
switch(Config-If-Vlan1)#no shutdown
switch(Config-If-Vlan1)#exit
switch(Config)#exit

第三步:创建 vlan100 和 vlan200。

switch(Config)#
switch(Config)#vlan 100
switch(Config-Vlan100)#exit
switch(Config)#vlan 200
switch(Config-Vlan200)#exit
switch(Config)#

验证配置:

switch#show vlan

VLAN	Name	Type	Media	Ports	
1	default	Static	ENET	Ethernet0/0/1	Ethernet0/0/2
				Ethernet0/0/3	Ethernet0/0/4
				Ethernet0/0/5	Ethernet0/0/6
				Ethernet0/0/7	Ethernet0/0/8
				Ethernet0/0/9	Ethernet0/0/10
				Ethernet0/0/11	Ethernet0/0/12
				Ethernet0/0/13	Ethernet0/0/14
				Ethernet0/0/15	Ethernet0/0/16
				Ethernet0/0/17	Ethernet0/0/18
				Ethernet0/0/19	Ethernet0/0/20
				Ethernet0/0/21	Ethernet0/0/22

				Ethernet0/0/23	Ethernet0/0/24
100	VLAN0100	Static	ENET	！已经创建了 vlan100，vlan100 中没有端口	
200	VLAN0200	Static	ENET	！已经创建了 vlan200，vlan200 中没有端口	

第四步：给 vlan100 和 vlan200 添加端口。

```
switch(Config)#vlan 100                                  ！进入 vlan 100
switch(Config-Vlan100)#switchport interface ethernet 0/0/1-8    ！给 vlan100 加入端口 1-8
Set the port Ethernet0/0/1 access vlan 100 successfully
Set the port Ethernet0/0/2 access vlan 100 successfully
Set the port Ethernet0/0/3 access vlan 100 successfully
Set the port Ethernet0/0/4 access vlan 100 successfully
Set the port Ethernet0/0/5 access vlan 100 successfully
Set the port Ethernet0/0/6 access vlan 100 successfully
Set the port Ethernet0/0/7 access vlan 100 successfully
Set the port Ethernet0/0/8 access vlan 100 successfully
switch(Config-Vlan100)#exit
switch(Config)#vlan 200                                  ！进入 vlan 200
switch(Config-Vlan200)#switchport interface ethernet 0/0/9-16   ！给 vlan200 加入端口 9-16
Set the port Ethernet0/0/9 access vlan 200 successfully
Set the port Ethernet0/0/10 access vlan 200 successfully
Set the port Ethernet0/0/11 access vlan 200 successfully
Set the port Ethernet0/0/12 access vlan 200 successfully
Set the port Ethernet0/0/13 access vlan 200 successfully
Set the port Ethernet0/0/14 access vlan 200 successfully
Set the port Ethernet0/0/15 access vlan 200 successfully
Set the port Ethernet0/0/16 access vlan 200 successfully
switch(Config-Vlan200)#exit
```

验证配置：

switch#show vlan

VLAN	Name	Type	Media	Ports	
1	default	Static	ENET	Ethernet0/0/17	Ethernet0/0/18
				Ethernet0/0/19	Ethernet0/0/20
				Ethernet0/0/21	Ethernet0/0/22
				Ethernet0/0/23	Ethernet0/0/24
100	VLAN0100	Static	ENET	Ethernet0/0/1	Ethernet0/0/2
				Ethernet0/0/3	Ethernet0/0/4
				Ethernet0/0/5	Ethernet0/0/6
				Ethernet0/0/7	Ethernet0/0/8
200	VLAN0200	Static	ENET	Ethernet0/0/9	Ethernet0/0/10
				Ethernet0/0/11	Ethernet0/0/12
				Ethernet0/0/13	Ethernet0/0/14
				Ethernet0/0/15	Ethernet0/0/16

第五步：验证实验。

实验结果如表 1-9 所示。

表 1-9 设备 Ping 的结果

PC1 位置	PC2 位置	动作	结果
1~8 端口		PC1 ping 192.168.2.11	不通
9~16 端口		PC1 ping 192.168.2.11	不通
17~24 端口		PC1 ping 192.168.2.11	通
1~8 端口	1~8 端口	PC1 ping PC2	通
1~8 端口	9~16 端口	PC1 ping PC2	不通
1~8 端口	17~24 端口	PC1 ping PC2	不通

七、注意事项和排错

1. 默认情况下，交换机所有端口都属于 vlan1，因此我们通常把 vlan1 作为交换机的管理 vlan，因此 vlan1 接口的 IP 地址就是交换机的管理地址。

2. 在 DCS-3926S 中，一个普通端口只属于一个 vlan。

八、配置序列

Switch#Show run
Current configuration:
!
　　hostname switch
!
Vlan 1
　　vlan 1
!
Vlan 100
　　vlan 100
!
Vlan 200
　　vlan 200
!
Interface Ethernet0/0/1
　　switchport access vlan 100
!
Interface Ethernet0/0/2
　　switchport access vlan 100
!
Interface Ethernet0/0/3
　　switchport access vlan 100
!

Interface Ethernet0/0/4
　　switchport access vlan 100
!
Interface Ethernet0/0/5
　　switchport access vlan 100
!
Interface Ethernet0/0/6
　　switchport access vlan 100
!
Interface Ethernet0/0/7
　　switchport access vlan 100
!
Interface Ethernet0/0/8
　　switchport access vlan 100
!
Interface Ethernet0/0/9
　　switchport access vlan 200
!
Interface Ethernet0/0/10
　　switchport access vlan 200
!
Interface Ethernet0/0/11
　　switchport access vlan 200
!
Interface Ethernet0/0/12
　　switchport access vlan 200
!
Interface Ethernet0/0/13
　　switchport access vlan 200
!
Interface Ethernet0/0/14
　　switchport access vlan 200
!
Interface Ethernet0/0/15
　　switchport access vlan 200
!
Interface Ethernet0/0/16
　　switchport access vlan 200
!
Interface Ethernet0/0/17
!
Interface Ethernet0/0/18
!
!

Interface Ethernet0/0/24
!
switch#

九、思考题

1. 怎样取消一个VLAN？
2. 怎样取消一个VLAN中的某些端口？

十、课后练习

请给交换机划分 3 个 VLAN，验证 VLAN 实验，VLAN 成员表如表 1-10 所示。

表 1-10　VLAN 划分

VLAN	端口成员
10	1~6
20	7~12
30	13~16

十一、相关配置命令详解

no name

命令：name <vlan-name>。

功能：为 VLAN 指定名称，VLAN 的名称是对该 VLAN 一个描述性字符串；本命令的 no 操作为删除 VLAN 的名称。

参数：<vlan-name>为指定的 VLAN 名称字符串。

命令模式：VLAN 配置模式。

默认情况：VLAN 默认 VLAN 名称为 VLANXXX，其中 XXX 为 VID。

使用指南：交换机提供为不同的 VLAN 指定名称的功能，有助于用户记忆 VLAN，方便管理。

举例：为 VLAN100 指定名称为 TestVlan。

Switch(Config-Vlan100)#name TestVlan

vlan

命令：vlan <vlan-id>。
　　　　no vlan <vlan-id>。

功能：创建 VLAN 并且进入 VLAN 配置模式，在 VLAN 模式中，用户可以配置 VLAN 名称和为该 VLAN 分配交换机端口；本命令的 no 操作为删除指定的 VLAN。

参数：<vlan-id>为要创建/删除的 VLAN 的 VID，取值范围为 1~4094。

命令模式：全局配置模式。

默认情况：交换机默认只有 VLAN1。

使用指南：VLAN1 为交换机的默认 VLAN，用户不能配置和删除 VLAN1。允许配置 VLAN 的总共数量为 255 个。另需要提醒的是不能使用本命令删除通过 GVRP 学习到的动态 VLAN。

举例：创建 VLAN100，并且进入 VLAN100 的配置模式。

Switch(Config)#vlan 100
Switch(Config-Vlan100)#

switchport interface

命令：switchport interface <interface-list>。
　　　no switchport interface <interface-list>。

功能：给 VLAN 分配以太网端口的命令；本命令的 no 操作为删除指定 VLAN 内的一个或一组端口。

参数：<interface-list>要添加或者删除的端口的列表，支持";"、"-"，如：ethernet 0/0/1;2;5 或 ethernet 0/0/1-6;8。

命令模式：VLAN 配置模式。

默认情况：新建立的 VLAN 默认不包含任何端口。

使用指南：Access 端口为普通端口，可以加入 VLAN，但同时只允许加入一个 VLAN。

举例：为 VLAN100 分配百兆以太网端口 1、3、4~7、8。

Switch(Config-Vlan100)#switchport interface ethernet 0/0/1;3;4-7;8

switchport mode

命令：switchport mode {trunk|access}。

功能：设置交换机的端口为 access 模式或者 trunk 模式。

参数：trunk 表示端口允许通过多个 VLAN 的流量；access 为端口只能属于一个 VLAN。

命令模式：端口配置模式。

默认情况：端口默认为 Access 模式。

使用指南：工作在 trunk mode 下的端口称为 Trunk 端口，Trunk 端口可以通过多个 VLAN 的流量，通过 Trunk 端口之间的互联，可以实现不同交换机上的相同 VLAN 的互通；工作在 access mode 下的端口称为 Access 端口，Access 端口可以分配给一个 VLAN，并且同时只能分配给一个 VLAN。

注意在 Trunk 端口不允许 802.1X 认证。

举例：将端口 5 设置为 trunk 模式，端口 8 设置为 access 模式。

Switch(Config)#interface ethernet 0/0/5
Switch(Config-ethernet0/0/5)#switchport mode trunk
Switch(Config-ethernet0/0/5)#exit
Switch(Config)#interface ethernet 0/0/8
Switch(Config-ethernet0/0/8)#switchport mode access
Switch(Config-ethernet0/0/8)#exit

实验八　跨交换机相同 VLAN 间通信

一、实验目的

1. 了解 IEEE802.1q 的实现方法，掌握跨二层交换机相同 VLAN 间通信的调试方法。
2. 了解交换机接口的 trunk 模式和 access 模式。
3. 了解交换机的 tagged 端口和 untagged 端口的区别。

二、应用环境

教学楼有两层，分别是一年级、二年级，每个楼层都有一台交换机满足老师上网需求；每个年级都有语文教研组和数学教研组；两个年级的语文教研组的计算机可以互相访问；两个年级的数学教研组的计算机可以互相访问；语文教研组和数学教研组之间不可以自由访问；通过划分VLAN使得语文教研组和数学教研组之间不可以自由访问；使用802.1Q进行跨交换机的VLAN。

三、实验设备

1. DCS-3926S 交换机　　2 台。
2. PC 机　　2 台。
3. Console 线　　1 根。
4. 直通网线　　2 根。

四、实验拓扑

该实验拓扑结构如图 1-33 所示。

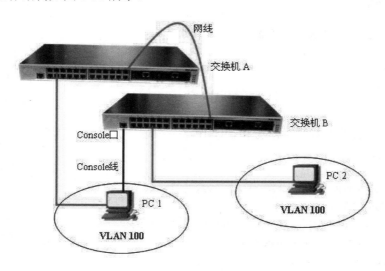

图 1-33　实验拓扑图

五、实验要求

在交换机 A 和交换机 B 上分别划分两个基于端口的 VLAN：VLAN100、VLAN200，如表 1-11 所示。

表 1-11 VLAN 成员表

VLAN	端口成员
100	1~8
200	9~16
Trunk	24

使得交换机之间 VLAN100 的成员能够互相访问，VLAN200 的成员能够互相访问；VLAN100 和 VLAN200 成员之间不能互相访问。

PC1 和 PC2 的网络设置如表 1-12 所示。

表 1-12 设备的 IP 地址分配

设备	IP 地址	MASK 子网掩码
交换机 A	192.168.2.11	255.255.255.0
交换机 B	192.168.2.12	255.255.255.0
PC1	192.168.2.101	255.255.255.0
PC2	192.168.2.102	255.255.255.0

PC1、PC2 分别接在不同交换机 VLAN100 的成员端口 1~8 上，两台 PC 互相可以 ping 通；PC1、PC2 分别接在不同交换机 VLAN 的成员端口 9~16 上，两台 PC 互相可以 ping 通；PC1 和 PC2 接在不同 VLAN 的成员端口上则互相 ping 不通。

若实验结果和理论相符，则本实验完成。

六、实验步骤

第一步：交换机恢复出厂设置。

switch#set default
switch#write
switch#reload

第二步：给交换机设置标示符和管理 IP。
交换机 A：

switch(Config)#hostname switchA
switchA(Config)#interface vlan 1
switchA(Config-If-Vlan1)#ip address 192.168.2.11 255.255.255.0
switchA(Config-If-Vlan1)#no shutdown
switchA(Config-If-Vlan1)#exit
switchA(Config)#

交换机 B：

switch(Config)#hostname switchB
switchB(Config)#interface vlan 1
switchB(Config-If-Vlan1)#ip address 192.168.2.12 255.255.255.0
switchB(Config-If-Vlan1)#no shutdown
switchB(Config-If-Vlan1)#exit
switchB(Config)#

第三步：在交换机中创建 vlan100 和 vlan200，并添加端口。

交换机 A：

switchA(Config)#vlan 100
switchA(Config-Vlan100)#
switchA(Config-Vlan100)#switchport interface ethernet 0/0/1-8
switchA(Config-Vlan100)#exit
switchA(Config)#vlan 200
switchA(Config-Vlan200)#switchport interface ethernet 0/0/9-16
switchA(Config-Vlan200)#exit
switchA(Config)#

验证配置：

switchA#show vlan

VLAN	Name	Type	Media	Ports	
1	default	Static	ENET	Ethernet0/0/17	Ethernet0/0/18
				Ethernet0/0/19	Ethernet0/0/20
				Ethernet0/0/21	Ethernet0/0/22
				Ethernet0/0/23	Ethernet0/0/24
100	VLAN0100	Static	ENET	Ethernet0/0/1	Ethernet0/0/2
				Ethernet0/0/3	Ethernet0/0/4
				Ethernet0/0/5	Ethernet0/0/6
				Ethernet0/0/7	Ethernet0/0/8
200	VLAN0200	Static	ENET	Ethernet0/0/9	Ethernet0/0/10
				Ethernet0/0/11	Ethernet0/0/12
				Ethernet0/0/13	Ethernet0/0/14
				Ethernet0/0/15	Ethernet0/0/16

switchA#

交换机 B：
配置与交换机 A 一样。

第四步：设置交换机 trunk 端口。

交换机 A：

switchA(Config)#interface ethernet 0/0/24
switchA(Config-Ethernet0/0/24)#switchport mode trunk

```
Set the port Ethernet0/0/24 mode TRUNK successfully
switchA(Config-Ethernet0/0/24)#switchport trunk allowed vlan all
set the port Ethernet0/0/24 allowed vlan successfully
switchA(Config-Ethernet0/0/24)#exit
switchA(Config)#
```

验证配置：

```
switchA#show vlan
```

VLAN	Name	Type	Media	Ports	
1	default	Static	ENET	Ethernet0/0/17	Ethernet0/0/18
				Ethernet0/0/19	Ethernet0/0/20
				Ethernet0/0/21	Ethernet0/0/22
				Ethernet0/0/23	Ethernet0/0/24(T)
100	VLAN0100	Static	ENET	Ethernet0/0/1	Ethernet0/0/2
				Ethernet0/0/3	Ethernet0/0/4
				Ethernet0/0/5	Ethernet0/0/6
				Ethernet0/0/7	Ethernet0/0/8
				Ethernet0/0/24(T)	
200	VLAN0200	Static	ENET	Ethernet0/0/9	Ethernet0/0/10
				Ethernet0/0/11	Ethernet0/0/12
				Ethernet0/0/13	Ethernet0/0/14
				Ethernet0/0/15	Ethernet0/0/16
				Ethernet0/0/24(T)	

```
switchA#
```

24 口已经出现在 vlan1、vlan100 和 vlan200 中，并且 24 口不是一个普通端口，是 tagged 端口。

交换机 B：

配置与交换机 A 一样。

第五步：验证实验。

交换机 A ping 交换机 B：

```
switchA#ping 192.168.2.12
Type ^c to abort.
Sending 5 56-byte ICMP Echos to 192.168.2.12, timeout is 2 seconds.
!!!!!
Success rate is 100 percent (5/5), round-trip min/avg/max = 1/1/1 ms
switchA#
```

表明交换机之前的 trunk 链路已经成功建立。按表 1-13 验证，PC1 连接在交换机 A 上，PC2 连接在交换机 B 上。

表 1-13　Ping 命令的结果

PC1 位置	PC2 位置	动作	结果
1~8 端口		PC1ping 交换机 B	不通
9~16 端口		PC1ping 交换机 B	不通
17~24 端口		PC1ping 交换机 B	通
1~8 端口	1~8 端口	PC1 ping PC2	通
1~8 端口	9~16 端口	PC1 ping PC2	不通

七、注意事项和排错

1. 取消一个 vlan 可以使用 "no vlan"。

2. 取消 vlan 的某个端口可以在 vlan 模式下使用 "no switchport interface ethernet 0/0/x"。

3. 当使用 "switchport trunk allowed vlan all" 命令后，所有以后创建的 vlan 中都会自动添加 trunk 口为成员端口。

八、配置序列

switchA#show run
Current configuration:
!
  hostname switchA
!
Vlan 1
  vlan 1
!
Vlan 100
  vlan 100
!
Vlan 200
  vlan 200
!
!
Interface Ethernet0/0/1
  switchport access vlan 100
!
Interface Ethernet0/0/2
  switchport access vlan 100
!
Interface Ethernet0/0/3
  switchport access vlan 100
!
Interface Ethernet0/0/4
  switchport access vlan 100

!
Interface Ethernet0/0/5
    switchport access vlan 100
!
Interface Ethernet0/0/6
    switchport access vlan 100
!
Interface Ethernet0/0/7
    switchport access vlan 100
!
Interface Ethernet0/0/8
    switchport access vlan 100
!
Interface Ethernet0/0/9
    switchport access vlan 200
!
Interface Ethernet0/0/10
    switchport access vlan 200
!
Interface Ethernet0/0/11
    switchport access vlan 200
!
Interface Ethernet0/0/12
    switchport access vlan 200
!
Interface Ethernet0/0/13
    switchport access vlan 200
!
Interface Ethernet0/0/14
    switchport access vlan 200
!
Interface Ethernet0/0/15
    switchport access vlan 200
!
Interface Ethernet0/0/16
    switchport access vlan 200
!
Interface Ethernet0/0/17
!
Interface Ethernet0/0/18
!
Interface Ethernet0/0/19
!
Interface Ethernet0/0/20

```
!
Interface Ethernet0/0/21
!
Interface Ethernet0/0/22
!
Interface Ethernet0/0/23
!
Interface Ethernet0/0/24
    switchport mode trunk
!
interface Vlan1
    interface vlan 1
    ip address 192.168.2.11 255.255.255.0
!
switchA#
```

九、思考题

Trunk、access、tagged 和 untagged 这几个专业术语的关联与区别是什么？

十、相关配置命令详解

switchport access vlan

命令：switchport access vlan <vlan-id>。

no switchport access vlan。

功能：将当前 Access 端口加入到指定 VLAN；本命令 no 操作为将当前端口从 VLAN 里删除。

参数：<vlan-id>为当前端口要加入的 vlanVID，取值范围为 1~4094。

命令模式：端口配置模式。

默认情况：所有端口默认属于 VLAN1。

使用指南：只有属于 Access mode 的端口才能加入到指定的 VLAN 中，并且 Access 端口同时只能加入到一个 VLAN 里去。

举例：设置某 Access 端口加入 VLAN100。

```
Switch(Config)#interface ethernet 0/0/8
Switch(Config-ethernet0/0/8)#switchport mode access
Switch(Config-ethernet0/0/8)#switchport access vlan 100
Switch(Config-ethernet0/0/8)#exit
```

switchport mode

命令：switchport mode {trunk|access}。

功能：设置交换机的端口为 access 模式或者 trunk 模式。

参数：trunk 表示端口允许通过多个 VLAN 的流量；access 为端口只能属于一个 VLAN。

命令模式：端口配置模式。

默认情况：端口默认为 Access 模式。

使用指南：工作在 trunk mode 下的端口称为 Trunk 端口，Trunk 端口可以通过多个 VLAN 的流量，通过 Trunk 端口之间的互联，可以实现不同交换机上的相同 VLAN 的互通；工作在 access mode 下的端口称为 Access 端口，Access 端口可以分配给一个 VLAN，并且同时只能分配给一个 VLAN。

注意在 Trunk 端口不允许 802.1X 认证。

举例：将端口 5 设置为 trunk 模式。

Switch(Config)#interface ethernet 0/0/5
Switch(Config-ethernet0/0/5)#switchport mode trunk
Switch(Config-ethernet0/0/5)#exit

端口 8 设置为 access 模式(删除 Trunk 的方法)。

Switch(Config)#interface ethernet 0/0/8
Switch(Config-ethernet0/0/8)#switchport mode access
Switch(Config-ethernet0/0/8)#exit

switchport trunk allowed vlan

命令：switchport trunk allowed vlan {<vlan-list>|all}。

　　　　no switchport trunk allowed vlan。

功能：设置 Trunk 端口允许通过 VLAN；本命令的 no 操作为恢复默认情况。

参数：<vlan-list>为允许在该 Trunk 端口上通过的 VLAN 列表；all 关键字表示允许该 Trunk 端口通过所有 VLAN 的流量。

命令模式：端口配置模式。

默认情况：Trunk 端口默认允许通过所有 VLAN。

使用指南：用户可以通过本命令设置哪些 VLAN 的流量通过 Trunk 端口，没有包含的 VLAN 流量则被禁止。

举例：设置 Trunk 端口允许通过 VLAN1、3、5~20 的流量。

Switch(Config)#interface ethernet 0/0/5
Switch(Config-ethernet0/0/5)#switchport mode trunk
Switch(Config-ethernet0/0/5)#switchport trunk allowed vlan 1;3;5-20
Switch(Config-ethernet0/0/5)#exit

switchport trunk native vlan

命令：switchport trunk native vlan <vlan-id>。

　　　　no switchport trunk native vlan。

功能：设置 Trunk 端口的 PVID；本命令的 no 操作为恢复默认值。

参数：<vlan-id>为 Trunk 端口的 PVID。

命令模式：端口配置模式。

默认情况：Trunk 端口默认的 PVID 为 1。

使用指南：在 802.1Q 中定义了 PVID 这个概念。Trunk 端口的 PVID 的作用是当一个 untagged 的帧进入 Trunk 端口，端口会对这个 untagged 帧打上带有本命令设置的 native PVID 的 tag 标记，用于 VLAN 的转发。

举例：设置某 Trunk 端口的 native vlan 为 100。

Switch(Config)#interface ethernet 0/0/5
Switch(Config-ethernet0/0/5)#switchport mode trunk
Switch(Config-ethernet0/0/5)#switchport trunk native vlan 100
Switch(Config-ethernet0/0/5)#exit

实训二 跨交换机相同 VLAN 间通信

一、实训设备

(1) DCS-3926S 交换机   2 台。
(2) PC 机   2 台。
(3) 直通网线   3 根。

二、实训拓扑

该实验拓扑结构如图 1-34 所示。

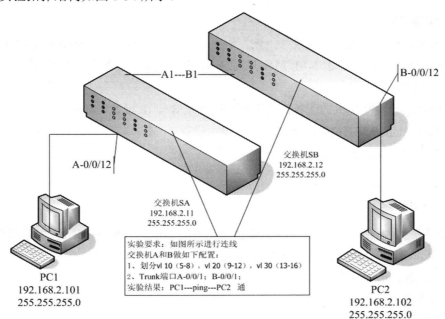

图 1-34　实验拓扑图

按照图 1-34 所示连线后，PC1 和 PC2 的网络设置如表 1-14 所示。

表 1-14 设备的 IP 地址分配

设备	IP 地址	Mask
交换机 A	192.168.2.11	255.255.255.0
交换机 B	192.168.2.12	255.255.255.0
A-PC1	192.168.2.101	255.255.255.0
B-PC2	192.168.2.102	255.255.255.0

PC1、PC2 接在 VLAN10 的成员端口 5~8 上，两台 PC 互相可以 ping 通；

PC1、PC2 接在 VLAN20 的成员端口 9~12 上，两台 PC 互相可以 ping 通；

PC1、PC2 接在 VLAN30 的成员端口 13~16 上，两台 PC 互相可以 ping 通；

PC1 接在 VLAN10 的成员端口 5~8 上，PC2 接在 VLAN20 的成员端口 9~12 上，则互相 ping 不通。

若实验结果和理论相符，则本实验完成。

验证实验结果，填写表 1-15。

表 1-15 实验结果对照表

PC1 位置	PC2 位置	动作	结果
SA　Vl10	SB　vl10	PC1 ping PC2	
SA　Vl10	SB　vl10	PC1 ping PC2	
SA　Vl10		PC1 ping SA	
SA　Vl10		PC1 ping SB	
		SA ping SB	

详细实验步骤说明：

(1) 按照实验要求选择对应设备，恢复出厂设置，做常规设置 lachhowww4.2A。

(2) 按照实验要求如图 3-14 所示进行正确连线，注意交换机级联位置 pc1-A12，PC2-B12，A1-B1。

(3) 按要求设置交换机的配置：交换机的地址及管理 IP 设置，vlan 划分：vlan10(5-8)，vlan20(9-12)，vlan30(13-16)以及级联端口的模式设置 Trunk。

(4) 设置测试主机的 IP 地址。

PC1　192.168.2.101/24
PC2　192.168.2.102/24

(5) 联通测试，使用 Ping 命令测试除自己以外的其他网络地址的连通情况，结果与实验要求结果作对比，符合实际情况，实验成功。

(6) PC1 ping PC2　通。

详细配置答案参考：

第一步：交换机恢复出厂设置。

switch#set default
switch#write
switch#reload
switch#la ch

第二步：给交换机设置标示符和管理 IP。
交换机 A：

switch(Config)#hostname ***A
switchA(Config)#interface vlan 1
switchA(Config-If-Vlan1)#ip address 192.168.2.11 255.255.255.0
switchA(Config-If-Vlan1)#no shutdown
switchA(Config-If-Vlan1)#exit
switchA(Config)#

交换机 B：

switch(Config)#hostname ***B
switchB(Config)#interface vlan 1
switchB(Config-If-Vlan1)#ip address 192.168.2.12 255.255.255.0
switchB(Config-If-Vlan1)#no shutdown
switchB(Config-If-Vlan1)#exit
switchB(Config)#

第三步：在交换机中创建 vlan100 和 vlan200，并添加端口。
交换机 A：

switchA(Config)#vlan 100
switchA(Config-Vlan100)#switchport interface ethernet 0/0/1-8
switchA(Config-Vlan100)#exit
switchA(Config)#vlan 200
switchA(Config-Vlan200)#switchport interface ethernet 0/0/9-16
switchA(Config-Vlan200)#exit
switchA(Config)#

验证配置：

switchA#show vlan

交换机 B：
配置与交换机 A 一样。
第四步：设置交换机 trunk 端口。
交换机 A：

switchA(Config)#interface ethernet 0/0/24
switchA(Config-Ethernet0/0/24)#switchport mode trunk

switchA(Config-Ethernet0/0/24)#switchport trunk allowed vlan all
switchA(Config-Ethernet0/0/24)#exit
switchA(Config)#

验证配置:

switchA#show vlan

24 口已经出现在 vlan1、vlan100 和 vlan200 中,并且 24 口不是一个普通端口,是 tagged 端口。

交换机 B:
配置与交换机 A 一样。
第五步:验证实验。
交换机 A ping 交换机 B:

switchA#ping 192.168.2.12
Type ^c to abort.
Sending 5 56-byte ICMP Echos to 192.168.2.12, timeout is 2 seconds.
!!!!!
Success rate is 100 percent (5/5), round-trip min/avg/max = 1/1/1 ms
switchA#

表明交换机之前的 trunk 链路已经成功建立。
实验验证结果如表 1-16 所示。

表 1-16 实验结果对照表

PC1 位置	PC2 位置	动作	结果
SA Vl10	SB vl10	PC1 ping PC2	通
SA Vl10	SB vl10	PC1 ping PC2	不通
SA Vl10		PC1 ping SA	不通
SA Vl10		PC1 ping SB	不通
		SA ping SB	通

实验九 交换机 MAC 与 IP 的绑定

一、实验目的

1. 了解什么情况下需要 MAC 与 IP 绑定。
2. 了解如何在接入交换机上配置 MAC 与 IP 的绑定。

二、应用环境

学校机房或者网吧等需要固定 IP 地址上网的场所，为了防止用户任意修改 IP 地址，造成 IP 地址冲突，可以使用 MAC 与 IP 绑定技术。将 MAC、IP 和端口绑定在一起，使用户不能随便修改 IP 地址，不能随便更改接入端口，从而使内部网络从管理上更加完善。使用交换机的 AM 功能可以做到 MAC 和 IP 的绑定，AM 全称为 access management，访问管理，它利用收到数据报文的信息，譬如源 IP 地址和源 MAC，与配置硬件地址池相比较，如果找到则转发，否则丢弃。

三、实验设备

1. DCS-3926S 交换机　　1 台。
2. PC 机　　2 台。
3. Console 线　　1~2 根。
4. 直通网线若干。

四、实验拓扑

该实验拓扑结构如图 1-35 所示。

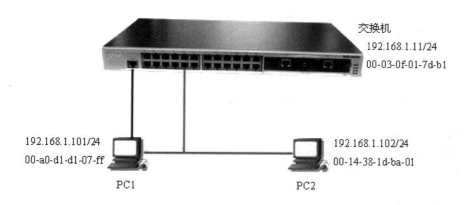

图 1-35　实验拓扑图

五、实验要求

1. 交换机 IP 地址为 192.168.2.11/24，PC1 的地址为 192.168.2.101/24；PC2 的地址为 192.168.2.102/24。
2. 在交换机 0/0/1 端口上将 PC1 的 IP、MAC 与端口绑定。
3. PC1 在 0/0/1 上 ping 交换机的 IP，检验理论是否和实验一致。
4. PC2 在 0/0/1 上 ping 交换机的 IP，检验理论是否和实验一致。
5. PC1 和 PC2 在其他端口上 ping 交换机的 IP，检验理论是否和实验一致。

六、实验步骤

第一步：得到 PC1 主机的 mac 地址。

Microsoft Windows XP [版本 5.1.2600]
(C) 版权所有 1985-2001 Microsoft Corp.
C:\>ipconfig/all
Windows IP Configuration

   Host Name . . . . : xuxp
   Primary Dns Suffix . . . . : digitalchina.com
   Node Type . . . . : Broadcast
   IP Routing Enabled. . . . . : No
   WINS Proxy Enabled. . . . . : No

Ethernet adapter 本地连接:

   Connection-specific DNS Suffix . :
   Description . . . . : Intel(R) PRO/100 VE Network Connection
   Physical Address. . . . . : 00-A0-D1-D1-07-FF
   Dhcp Enabled. . . . . : Yes
   Autoconfiguration Enabled . . . . : Yes
   Autoconfiguration IP Address. . . : 169.254.27.232
   Subnet Mask . . . . : 255.255.0.0
   Default Gateway . . . . :

C:\>

我们得到了 PC1 主机的 mac 地址为：00-A0-D1-D1-07-FF。

第二步：交换机全部恢复出厂设置，配置交换机的 IP 地址。

```
switch(Config)#interface vlan 1
switch(Config-If-Vlan1)#ip address 192.168.2.11 255.255.255.0
switch(Config-If-Vlan1)#no shut
switch(Config-If-Vlan1)#exit
switch(Config)#
```

第三步：使能 am 功能。

```
switch(Config)#am enable
switch(Config)#interface ethernet 0/0/1
switch(Config-Ethernet0/0/1)#am mac-ip-pool 00-A0-D1-D1-07-FF 192.168.2.101
switch(Config-Ethernet0/0/1)#exit
```

验证配置：

```
switch#show am
Am is enabled
Interface Ethernet0/0/1
am mac-ip-pool 00-A0-D1-D1-07-FF 192.168.2.101 USER_CONFIG
```

第四步：解锁其他端口。

Switch(Config)#interface ethernet 0/0/2
Switch(Config-Ethernet0/0/2)#no am port
Switch(Config)#interface ethernet 0/0/3-20
Switch(Config-Ethernet0/0/3-20)#no am port

第五步：使用 ping 命令验证。

验证结果如表 1-17 所示。

表 1-17　ping 命令结果

PC	端口	Ping	结果	原因
PC1	0/0/1	192.168.2.11	通	
PC1	0/0/7	192.168.2.11	通	
PC2	0/0/1	192.168.2.11	不通	
PC2	0/07	192.168.2.11	通	
PC1	0/0/21	192.168.2.11	不通	
PC2	0/0/21	192.168.2.11	不通	

七、注意事项和排错

1. AM 的默认动作是：拒绝通过(deny)，当 AM 使能的时候，AM 模块会拒绝所有的 IP 报文通过(只允许 IP 地址池内的成员源地址通过)，AM 禁止的时候，AM 会删除所有的地址池。

2. 对 AM，由于其硬件资源有限，每个 block(8 个端口)最多只能配置 256 条表项。

3. AM 资源要求用户配置的 IP 地址和 MAC 地址不能冲突，也就是说，同一个交换机上不同用户不允许出现相同的 IP 或 MAC 配置。

八、相关配置命令详解

am enable

命令：am enable。

　　　　no am enable。

功能：使能访问控制功能，在 am enable 时，AM 模块会拒绝所有的 IP 报文通过；no 命令禁止访问管理功能，并清空 IP 地址池和 MAC-IP 地址池。

命令模式：全局配置模式。

默认情况：am 访问控制不使能。

使用指南：在 AM 使能后，交换机禁止所有的 IP 报文，需要用户手动在各个接口上配置 IP 地址或 MAC-IP 地址才能使用户间互通；AM 禁止时，删除所有用户配置的地址。

举例：使能 AM。

Switch(Config)#am enable
am ip_pool

命令：am ip-pool <start_ip_address> [<num>]。

　　　　no am ip-pool<start_ip_address> [<num>]。

功能：创建一个 IP 地址段，放到地址池中。no 命令删除一个在地址池中已经配置的 IP 地址段。

参数：start_ip_address 是 IP 地址中某段地址范围的起始地址。num 是从 start_ip_address 开始的连续的地址数量，默认值是 1。

命令模式：物理接口配置模式。

默认情况：ip-pool 为空。

使用指南：用户调用此命令，自己配置 IP 地址池内容，允许相应的接口上相应的源 IP 通过。

举例：使能 AM 并允许接口 4 上源 IP 为 192.1.1.2~192.1.1.10 的 9 个用户通过。

Switch(Config)#am enable
Switch(Config)#interface Ethernet 0/0/4
Switch(Config-Ethernet0/0/4)#ip pool 192.1.1.2　9

am mac_ip_pool

命令：am mac-ip-pool <mac_address> <ip_address>。

　　　　no am mac-ip-pool <mac_address>< ip_address>。

功能：创建一个 MAC+IP 地址绑定，放到地址池中，或者删除一个在地址池中已经配置的 MAC+IP 地址绑定，MAC 和 IP 地址一一对应。

参数：

mac_address：源 MAC 地址格式为 HH-HH-HH-HH-HH-HH。

ip_address：源 IP 地址。32 位二进制数，用 4 个隔开的十进制数表示。

命令模式：物理接口配置模式。

默认配置：MAC+IP pool 为空。

使用指南：用户调用此命令，配置 MAC-IP 地址池内容，允许相应的接口上相应的源 MAC-IP 通过。

举例：使能 AM 并允许接口 4 上源 IP 为 192.1.1.2 源 MAC 是 00-01-10-22-33-10 的用户通过：

Switch(Config)#am enable
Switch(Config)#interface Ethernet 0/0/4
Switch(Config-Ethernet0/0/4)#mac-ip pool 00-01-10-22-33-10 192.1.1.2

am port

命令：am port。

　　　　no am port。

功能：打开或关闭物理接口上的 AM 功能。

命令模式：物理接口配置模式。

默认情况：AM 功能打开。

使用指南：用户可以关闭端口的 AM 功能，一般应用于上联口。

举例：关闭 0/1/1 端口的 AM 功能。

Switch(Config)#am enable
Switch(Config)#interface Ethernet 0/1/1
Switch(Config-Ethernet0/1/1)#no am port

no am all

命令：no am all {ip-pool|mac-ip-pool}。

功能：删除全部用户配置的 MAC-IP 地址池或者 IP 地址池。

ip-pool：IP 地址池。

mac-ip-pool：MAC-IP 地址池。

all：所有 IP 地址池或者 MAC 地址池。

命令模式：全局配置模式。

默认配置：无。

使用指南：用户调用此命令，清空 MAC-IP 地址池或者 IP 地址池内全部用户配置的地址。

举例：清空全部用户配置的 MAC-IP 地址。

Switch(Config)#no am enable all mac-ip-pool

show am

命令：

show am [interface <interfaceName>]。

功能：显示当前交换机配置的地址项。

Interface Name：物理接口名。

命令模式：全局配置模式。

默认配置：无。

使用指南：不指定访问接口的名字时，会显示所有的访问控制列表。

举例：

Switch#show am
am is enabled
Interface Ethernet0/0/10
    am mac-ip-pool 00-00-00-00-00-13 100.1.1.2 USER_CONFIG
    am mac-ip-pool 00-00-00-00-01-12 100.1.1.1 USER_CONFIG
Interface Ethernet0/0/1
   am ip-pool 10.1.1.1 8   USER_CONFIG

详细参数如表 1-18 所示。

表 1-18　参数说明对照解释

显示内容	解释
am is enabled	AM 是使能的
Am mac-ip-pool　00-00-00-00-00-13　100.1.1.2　USER_CONFIG	用户源 MAC=00-00-00-00-00-13，且同时源 IP=100.1.1.2 的用户才能通过，该项由用户配置
Am mac-ip-pool　00-00-00-00-01-12　100.1.1.1　USER_CONFIG	用户源 MAC=00-00-00-00-01-12，且同时源 IP=100.1.1.1 的用户才能通过，该项由用户配置
am ip-pool 10.1.1.1 8 USER_CONFIG	用户源 IP=10.1.1.1~10.1.1.8 的用户才能通过，该项由用户配置

实验十　生成树实验

一、实验目的

1. 了解生成树协议的作用。
2. 熟悉生成树协议的配置。

二、应用环境

交换机之间具有冗余链路本来是一件很好的事情，但是它有可能引起的问题比它能够解决的问题还要多。如果用户真的准备两条以上的路，就必然形成了一个环路，交换机并不知道如何处理环路，只是周而复始地转发帧，形成一个"死循环"，这个死循环会造成整个网络处于阻塞状态，导致网络瘫痪。采用生成树协议可以避免环路。

生成树协议的根本目的是将一个存在物理环路的交换网络变成一个没有环路的逻辑树形网络。IEEE802.1d 协议通过在交换机上运行一套复杂的算法 STA(spanning-treealgorithm)，使冗余端口置于"阻断状态"，使得接入网络的计算机在与其他计算机通信时，只有一条链路生效，而当这个链路出现故障无法使用时，IEEE802.1d 协议会重新计算网络链路，将处于"阻断状态"的端口重新打开，从而既保障了网络正常运转，又保证了冗余能力。

三、实验设备

1. DCS-3926S交换机　　2台。
2. PC机　　2台。
3. Console线　　1~2根。
4. 直通网线　　4~8根。

四、实验拓扑

该实验拓扑结构如图 1-36 所示。

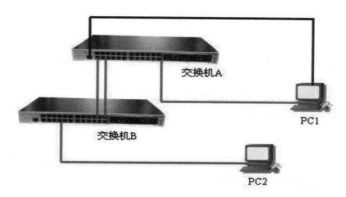

图 1-36 实验拓扑图

五、实验要求

该实验中各项配置如表 1-19 所示，连线如表 1-20 所示。

表 1-19　IP 地址设置

设备	IP	Mask
交换机 A	192.168.2.11	255.255.255.0
交换机 B	192.168.2.12	255.255.255.0
PC1	192.168.2.101	255.255.255.0
PC2	192.168.2.102	255.255.255.0

表 1-20　网线连接

交换机 A e0/0/1	交换机 B e0/0/3
交换机 A e0/0/2	交换机 B e0/0/4
PC1	交换机 A e0/0/24
PC2	交换机 B e0/0/23

如果生成树成功，则 PC1 可以 ping 通 PC2。

六、实验步骤

第一步：正确连接网线，恢复出厂设置之后，做初始配置。

交换机 A：

```
switch#config
switch(Config)#hostname switchA
switchA(Config)#interface vlan 1
switchA(Config-If-Vlan1)#ip address 192.168.2.11 255.255.255.0
switchA(Config-If-Vlan1)#no shutdown
switchA(Config-If-Vlan1)#exit
switchA(Config)#
```

交换机 B：

switch#config
switch(Config)#hostname switchB
switchB(Config)#interface vlan 1
switchB(Config-If-Vlan1)#ip address 192.168.2.12   255.255.255.0
switchB(Config-If-Vlan1)#no shutdown
switchB(Config-If-Vlan1)#exit
switchB(Config)#

第二步："PC1 ping PC2 -t"观察现象。

1. ping 不通。

2. 所有连接网线端口的绿灯很频繁地闪烁，表明该端口收发数据量很大，已经在交换机内部形成广播风暴。

第三步：在两台交换机中都启用生成树协议。

switchA(Config)#spanning-tree
MSTP is starting now, please wait...........
MSTP is enabled successfully.
switchA(Config)#
switchB(Config)#spanning-tree
MSTP is starting now, please wait...........
MSTP is enabled successfully.
switchB(Config)#

验证配置：

switchA#show spanning-tree
                      -- MSTP Bridge Config Info --
Standard   :   IEEE 802.1s
Bridge MAC   :   00:03:0f:01:25:28
Bridge Times :   Max Age 20, Hello Time 2, Forward Delay 15
Force Version: 3
######################### Instance 0 #########################
Self Bridge Id   : 32768 -   00:03:0f:01:25:28
Root Id   : this switch
Ext.RootPathCost : 0
Region Root Id   : this switch
Int.RootPathCost : 0
Root Port ID   : 0
Current port list in Instance 0:
Ethernet0/0/1 Ethernet0/0/2 (Total 2)
       PortName   ID   ExtRPC   IntRPC   State Role   DsgBridge   DsgPort
    -------------- ------- --------- --------- ---- ---- ------------------ -------
    Ethernet0/0/1 128.001   0   0   FWD DSGN 32768.00030f012528 128.001

```
          Ethernet0/0/2 128.002        0         0 FWD DSGN 32768.00030f012528 128.002
switchA#
switchB#show spanning-tree
                -- MSTP Bridge Config Info --

Standard          :   IEEE 802.1s
Bridge MAC        :   00:03:0f:01:7d:b0
Bridge Times :    Max Age 20, Hello Time 2, Forward Delay 15
Force Version:    3
######################## Instance 0 ########################
Self Bridge Id    : 32768 -    00:03:0f:01:7d:b0
Root Id           : 32768.00:03:0f:01:25:28
Ext.RootPathCost : 200000
Region Root Id    : this switch
Int.RootPathCost : 0
Root Port ID      : 128.4
Current port list in Instance 0:
Ethernet0/0/3 Ethernet0/0/4 Ethernet0/0/23 (Total 3)

   PortName       ID       ExtRPC    IntRPC   State Role    DsgBridge           DsgPort
  -------------- --------- --------- -------- --- ---- ------------------- -------
   Ethernet0/0/3  128.003    0          0     BLK ALTR   32768.00030f012528   128.002
   Ethernet0/0/4  128.004    0          0     FWD ROOT   32768.00030f012528   128.001
   Ethernet0/0/23 128.023    200000     0     FWD DSGN   32768.00030f017db0   128.023
```

从 show 中可以看出，交换机 A 是根交换机，交换机 B 的 4 端口是根端口。

第四步：继续使用"PC1 ping PC2 -t"观察现象。

1. 拔掉交换机 B 端口 4 的网线，观察现象，写在下方。
2. 再插上交换机 B 端口 4 的网线，观察现象，写在下方。

七、注意事项和排错

1. 如果想在交换机上运行 MSTP，首先必须在全局打开 MSTP 开关。在没有打开全局 MSTP 开关之前，打开端口的 MSTP 开关是不允许的。

2. MSTP 定时器参数之间是有相关性的，错误配置可能导致交换机不能正常工作。各定时器之间的关联关系为：

(1) $2 \times (Bridge\_Forward\_Delay - 1.0\ seconds) >= Bridge\_Max\_Age$。

(2) $Bridge\_Max\_Age >= 2 \times (Bridge\_Hello\_Time + 1.0\ seconds)$。

3. 用户在修改 MSTP 参数时，应该清楚所产生的各个拓扑。除了全局的基于网桥的参数配置外，其他的是基于各个实例的配置，在配置时一定要注意配置参数对应的实例是否正确。

4. DCS-3926S 交换机的端口 MSTP 功能与端口 MAC 绑定、802.1x 和设置端口为路由端口功能互斥。当端口已经配置 MAC 绑定、802.1x 或设置为路由端口时，无法在该端口启动 MSTP 功能。

八、配置序列

略。

九、课后练习

1. 使用 4 根网线连接两台交换机,观察根端口的选择,观察备份线路启用时的 debug 信息。

2. 使用"spanning-tree mode mstp"来进行上面的实验,体验备份链路启用和断开所需要的时间长短。

十、思考题

1. 生成树协议怎样选取根端口和指定端口?
2. MSTP 通过怎样的策略可以使备份链路实现快速启用?

十一、相关配置命令详解

spanning-tree

命令:spanning-tree。

no spanning-tree。

功能:在交换机的全局配置模式和端口配置模式下分别启动 MSTP 协议的命令;本命令的 no 操作为关闭 MSTP 协议。

参数:无。

命令模式:全局配置模式和端口配置模式。

默认情况:系统默认不运行 MSTP 协议。但如果在全局配置模式下启动了 MSTP 协议,所有的端口默认都打开 MSTP 协议。

使用指南:用户若要进行 MSTP 参数的配置,必须在全局模式下首先启动 MSTP 协议。

举例:在全局模式打开 MSTP,并且在端口 0/0/2 模式关闭 MSTP。

Switch(Config)#spanning-tree
Switch(Config)#interface ethernet 0/0/2
Switch(Config-Ethernet0/0/2)#no spanning-tree

实验十一 交换机链路聚合

一、实验目的

1. 了解链路聚合技术的使用场合。

2. 熟练掌握链路聚合技术的配置。

二、应用环境

两个实验室分别使用一台交换机提供20多个信息点，两个实验室的互通通过一根级联网线。每个实验室的信息点都是百兆到桌面。两个实验室之间的带宽也是100M，如果实验室之间需要大量传输数据，就会明显感觉带宽资源紧张。当楼层之间大量用户都希望以100M 传输数据时，楼层间的链路就呈现出了独木桥的状态，必然造成网络传输效率下降等后果。

解决这个问题的办法就是提高楼层主交换机之间的连接带宽，实现的办法可以是采用千兆端口替换原来的100M 端口进行互联，但这样无疑会增加组网的成本，需要更新端口模块，并且线缆也需要作进一步的升级。另一种相对经济的升级办法就是链路聚合技术。顾名思义，链路聚合是将几个链路作聚合处理，这几个链路必须是同时连接两个相同的设备，这样，当作了链路聚合之后就可以实现几个链路相加的带宽了。比如，我们可以将4个100M 链路使用链路聚合做成一个逻辑链路，这样在全双工条件下就可以达到800M 的带宽，即将近1000M 的带宽。这种方式比较经济，实现也相对容易。

三、实验设备

1. DCS-3926S 交换机　　2 台。
2. PC 机　　2 台。
3. Console 线　　1~2 根。
4. 直通网线　　4~8 根。

四、实验拓扑

交换机链路聚合的实验拓扑图如图 1-37 所示。

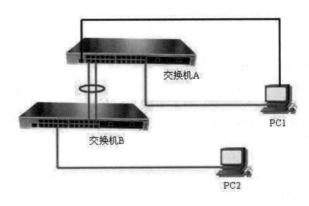

图 1-37　实验拓扑图

五、实验要求

设备 IP 配置如表 1-21 所示。

表 1-21　实验配置

设备	IP	Mask	端口
交换机 A	192.168.2.11	255.255.255.0	0/0/1-2 trunking
交换机 B	192.168.2.12	255.255.255.0	0/0/1-2 trunking
PC1	192.168.2.101	255.255.255.0	交换机 A0/0/23
PC2	192.168.2.102	255.255.255.0	交换机 B0/0/24

如果链路聚合成功，则 PC1 可以 ping 通 PC2。

六、实验步骤

第一步：正确连接网线，交换机全部恢复出厂设置，进行初始配置，避免广播风暴出现。

交换机 A：

switch#config
switch(Config)#hostname switchA
switchA(Config)#interface vlan 1
switchA(Config-If-Vlan1)#ip address 192.168.2.11 255.255.255.0
switchA(Config-If-Vlan1)#no shutdown
switchA(Config-If-Vlan1)#exit
switchA(Config)#spanning-tree
MSTP is starting now, please wait..........
MSTP is enabled successfully.
switchA(Config)#

交换机 B：

switch#config
switch(Config)#hostname switchB
switchB(Config)#interface vlan 1
switchB(Config-If-Vlan1)#ip address 192.168.2.12 255.255.255.0
switchB(Config-If-Vlan1)#no shutdown
switchB(Config-If-Vlan1)#exit
switchB(Config)#spanning-tree
MSTP is starting now, please wait..........
MSTP is enabled successfully.
switchB(Config)#

第二步：创建 port group。

交换机 A：

switchA(Config)#port-group 1
switchA(Config)#

验证配置：

switchA#show port-group detail
Sorted by the ports in the group 1:
----
switchA#show port-group brief
Port-group number : 1
Number of ports in port-group : 0   Maxports in port-channel = 8
Number of port-channels : 0   Max port-channels : 1
switchA#

交换机 B：

switchB(Config)#port-group 2
switchB(Config)#

第三步：手工生成链路聚合组(第三、四步任选其一操作)。
交换机 A：

switchA(Config)#interface ethernet 0/0/1-2
switchA(Config-Port-Range)#port-group 1 mode on
switchA(Config-Port-Range)#exit
switchA(Config)#interface port-channel 1
switchA(Config-If-Port-Channel1)#

验证配置：

switchA#show vlan

VLAN	Name	Type	Media	Ports	
1	default	Static	ENET	Ethernet0/0/3	Ethernet0/0/4
				Ethernet0/0/5	Ethernet0/0/6
				Ethernet0/0/7	Ethernet0/0/8
				Ethernet0/0/9	Ethernet0/0/10
				Ethernet0/0/11	Ethernet0/0/12
				Ethernet0/0/13	Ethernet0/0/14
				Ethernet0/0/15	Ethernet0/0/16
				Ethernet0/0/17	Ethernet0/0/18
				Ethernet0/0/19	Ethernet0/0/20
				Ethernet0/0/21	Ethernet0/0/22
				Ethernet0/0/23	Ethernet0/0/24
				Port-Channel1	
switchA#				! port-channel1 已经存在	

交换机 B：

switchB(Config)#int e 0/0/3-4
switchB(Config-Port-Range)#port-group 2 mode on
switchB(Config-Port-Range)#exit
switchB(Config)#interface port-channel 2
switchB(Config-If-Port-Channel2)#

验证配置：

switchB#show port-group brief
Port-group number : 2
Number of ports in port-group : 2   Maxports in port-channel = 8
Number of port-channels : 1   Max port-channels : 1
switchB#

第四步： LACP 动态生成链路聚合组(第三、四步任选其一操作)。

switchA(Config)#interface ethernet 0/0/1-2
switchA(Conifg-Port-Range)#port-group 1 mode active
switchA(Config)#interface port-channel 1
switchA(Config-If-Port-Channel1)#

验证配置：

switchA#show vlan

VLAN	Name	Type	Media	Ports	
1	default	Static	ENET	Ethernet0/0/3	Ethernet0/0/4
				Ethernet0/0/5	Ethernet0/0/6
				Ethernet0/0/7	Ethernet0/0/8
				Ethernet0/0/9	Ethernet0/0/10
				Ethernet0/0/11	Ethernet0/0/12
				Ethernet0/0/13	Ethernet0/0/14
				Ethernet0/0/15	Ethernet0/0/16
				Ethernet0/0/17	Ethernet0/0/18
				Ethernet0/0/19	Ethernet0/0/20
				Ethernet0/0/21	Ethernet0/0/22
				Ethernet0/0/23	Ethernet0/0/24
				Port-Channel1	

switchA#   ！port-channel1 已经存在

交换机 B：

switchB(Config)#interface ethernet 0/0/3-4
switchB(Conifg-Port-Range)#port-group 2 mode passive
switchB(Config)#interface port-channel 2

switchB(Config-If-Port-Channel2)#

验证配置：

switchB#show port-group brief
Port-group number : 2
Number of ports in port-group : 2   Maxports in port-channel = 8
Number of port-channels : 1   Max port-channels : 1
switchB#

第五步：如表 1-22 所示，使用 ping 命令验证，使用 PC1 ping PC2。

表 1-22 测试配置表

交换机 A	交换机 B	结果	原因
0/0/1 0/0/2	0/0/3 0/0/4	通	链路聚合组连接正确
0/0/1 0/0/2	0/0/3	通	拔掉交换机 B 端口 4 的网线，仍然可以通(需要一点时间)，此时用 show vlan 查看结果，port-channel 消失。只有一个端口连接时，没有必要再维持一个 port-channel
0/0/1 0/0/2	0/0/5 0/0/6	通	等候一小段时间后，仍然是通的。用 show vlan 看结果。此时把两台交换机的 spanning-tree 功能 disable 掉，这时候使用第三步和第四步的结果会不同。采用第四步的，将会形成环路

七、注意事项和排错

1. 为使 Port Channel 正常工作，Port Channel 的成员端口必须具备以下相同的属性：
(1) 端口均为全双工模式。
(2) 端口速率相同。
(3) 端口的类型必须一样，比如同为以太口或同为光纤口。
(4) 端口同为 Access 端口并且属于同一个 VLAN 或同为 Trunk 端口。
(5) 如果端口为 Trunk 端口，则其 Allowed VLAN 和 Native VLAN 属性也应该相同。

2. 支持任意两个交换机物理端口的汇聚，最大组数为 6 个，组内最多的端口数为 8 个。

3. 一些命令不能在 port-channel 上的端口使用，包括 arp、bandwidth、ip、ip-forward 等。

4. 在使用强制生成端口聚合组时，由于汇聚是手工配置触发的，如果由于端口的 VLAN 信息不一致导致汇聚失败的话，汇聚组一直会停留在没有汇聚的状态，必须通过往该 group 增加和删除端口来触发端口再次汇聚，如果 VLAN 信息还是不一致仍然不能汇聚成功，直到 VLAN 信息都一致并且有增加和删除端口触发汇聚的情况下端口才能汇聚成功。

5. 检查对端交换机的对应端口是否配置端口聚合组，且要查看配置方式是否相同，如果本端是手工方式则对端也应该配置成手工方式，如果本端是 LACP 动态生成则对端也应该是 LACP 动态生成，否则端口聚合组不能正常工作；还有一点要注意的是，如果两端收发的都是 LACP 协议，至少有一端是 Active 的，否则两端都不会发起 LACP 数据报。

6. port-channel 一旦形成之后，所有对应端口的设置只能在 port-channel 端口上进行。

7. LACP 必须和 Security 和 802.1X 的端口互斥，如果端口已经配置上述两种协议，就不允许被启用 LACP。

八、课后练习

九、相关配置命令详解

port-group

命令：port-group <port-group-number> [load-balance { src-mac|dst-mac | dst-src-mac | src-ip | dst-ip|dst-src-ip}]

no port-group <port-group-number> [load-balance]。

功能：新建一个 port group，并且设置该组的流量分担方式。如果没有指定流量分担方式则为设置默认的流量分担方式。该命令的 no 操作为删除该 group 或者恢复该组流量分担的默认值，输入 load-balance 表示恢复默认流量分担，否则为删除该组。

参数：<port-group-number>为 Port Channel 的组号，范围为 1~16，如果已经存在该组号则会报错。dst-mac 根据目的 MAC 进行流量分担；src-mac 根据源 MAC 地址进行流量分担；dst-src-mac 根据目的 MAC 和源 MAC 进行流量分担；dst-ip 根据目的 IP 地址进行流量分担；src-ip 根据源 IP 地址进行流量分担；dst-src-ip 根据目的 IP 和源 IP 进行流量分担。如果是修改流量分担方式，并且该 port-group 已经形成一个 port-channel，则这次修改的流量分担方式只有在下次再次汇聚时才会生效。

默认情况：默认交换机端口不属于 Port Channel，不启动 LACP 协议。

命令模式：交换机全局配置模式。

举例：新建一个 port group，并且采用默认的流量分担方式。

Switch(Config)#port-group 1

删除一个 port group：

Switch(Config)#no port-group 1

port-group mode

命令：port-group <port-group-number> mode {active|passive|on}。

no port-group <port-group-number>。

功能：将物理端口加入 Port Channel，该命令的 no 操作为将端口从 Port Channel 中去除。

参数：<port-group-number>为 Port Channel 的组号，范围为 1~16；active(0)启动端口的 LACP 协议，并设置为 Active 模式；passive(1)启动端口的 LACP 协议，并且设置为 Passive 模式；on(2)强制端口加入 Port Channel，不启动 LACP 协议。

命令模式：交换机端口配置模式。

默认情况：默认交换机端口不属于 Port Channel，不启动 LACP 协议。

使用指南：如果不存在该组则会先建立该组，然后再将端口加到组中。在一个 port-group 中所有的端口加入的模式必须一样，以第一个加入该组的端口模式为准。端口以 on 模式加入一个组是强制性的，所谓强制性的表示本端交换机端口汇聚不依赖对端的信息，只要在组中有两个以上的端口，并且这些端口的 VLAN 信息都一致，则组中的端口就能汇聚成功。端口以 Active 和 Passive 方式加入一个组是运行 LACP 协议的，但两端必须有一个组中的端口是以 Active 方式加入的，如果两端都是 Passive，端口永远都无法汇聚起来。

举例：在 Ethernet0/0/1 端口模式下，将本端口以 Active 模式加入 port-group 1。

Switch(Config-Ethernet0/0/1)#port-group 1 mode active

interface port-channel

命令：interface port-channel <port-channel-number>。

功能：进入汇聚端口配置模式。

命令模式：全局配置模式。

使用指南：进入汇聚端口模式下配置时，如果是对 gvrp、spanningtree 模块做配置则对汇聚端口生效，如果汇聚端口不存在，也就是说在端口没有汇聚起来时先提示错误信息，记录该用户配置操作，当端口真正汇聚起来以后恢复用户刚才对未形成汇聚端口的配置动作，注意只能恢复一次，如果因为某种原因汇聚组被拆散然后又汇聚起来，用户一开始的配置不能被恢复。如果是对其他模块做配置，比如做 shutdown、speed 配置，则是对该 port-channel 对应的 port-group 中的所有成员端口生效，起到一个群配的作用。

举例：进入 port-channel 1 配置模式。

Switch(Config)#interface port-channel 1
Switch(Config-If-Port-Channel1)#

show port-group

命令：show port-group [<port-group-number>] {brief | detail | load-balance | port | port-channel}。

参数：<port-group-number> 要显示的 Port Channel 的组号，范围为 1~16；brief 显示摘要信息；detail 显示详细信息；load-balance 显示流量分担信息；port 显示成员端口信息；port-channel 显示汇聚端口信息。

命令模式：特权配置模式。

使用指南：如果没有指定 port-group-number 则表示显示所有 port-group 的信息。

举例：将端口 0/0/1 和 0/0/2 加入 port-group 1 中，显示交换机的 port-group 1 的摘要信息。

Switch#show port-group 1 brief
Port-group number : 1
Number of ports in group : 2   Maxports = 8
Number of port-channels : 0   Max port-channels : 1

参数说明如表 1-23 所示。

表 1-23 显示信息表

显 示 内 容	解 释
Number of ports in group	在 port-group 中的端口数
Maxports	组中最大允许的端口数
Number of port-channels	是否已经汇聚成一个汇聚端口
Max port-channels	port-group 所能形成的最大汇聚端口数

1. 显示交换机的 port-group 1 的详细信息。

Switch# show port-group 1 detail
Sorted by the ports in the group 1:
----
port Ethernet0/0/1 :
both of the port and the agg attributes are not equal
the general information of the port are as follows:
portnumber: 1   actor_port_agg_id:0   partner_oper_sys:0x000000000000
partner_oper_key: 0x0001   actor_oper_port_key: 0x0101
mode of the port: ACTIVE   lacp_aware: enable
begin: FALSE   port_enabled: FALSE   lacp_ena: FALSE   ready_n: TRUE
the attributes of the port are as follows:
mac_type: ETH_TYPE   speed_type: ETH_SPEED_10M
duplex_type: FULL   port_type: ACCESS
the machine state and port state of the port are as the follow
mux_state: DETCH   rcvm_state: P_DIS   prm_state: NO_PER
actor_oper_port_state : L_A___F_
partner_oper_port_state: _TA___F_
port Ethernet0/0/2 :
both of the port and the agg attributes are not equal
the general information of the port are as follows:
portnumber: 2   actor_port_agg_id:0   partner_oper_sys:0x000000000000
partner_oper_key: 0x0002   actor_oper_port_key: 0x0102
mode of the port: ACTIVE   lacp_aware: enable
begin: FALSE   port_enabled: FALSE   lacp_ena: TRUE   ready_n: TRUE
the attributes of the port are as follows:
mac_type: ETH_TYPE   speed_type: ETH_SPEED_100M
duplex_type: FULL   port_type: ACCESS
the machine state and port state of the port are as the follow
mux_state: DETCH   rcvm_state: P_DIS   prm_state: NO_PER
actor_oper_port_state : L_A___F_
partner_oper_port_state: _TA___F_

详细参数如表 1-24 所示。

表 1-24 详细信息表

显 示 内 容	解　　释
portnumber	端口号
actor_port_agg_id	该端口加入的 channel 号，如果由于端口参数与 channel 的参数不一致，导致该端口不能进入 channel，显示为 0
partner_oper_sys	对端的 system ID
partner_oper_key	对端的 operational key
actor_oper_port_key	本端的 operational key
mode of the port	端口加入组的模式
mac_type	端口的类型，分标准以太网口和光纤分布式数据接口
speed_type	端口的 speed 类型，分 10M 和 100M 口
duplex_type	端口的双工类型，分全双工和半双工
port_type	端口的 vlan 属性，分 access 口和 trunk 口

2. 显示交换机的 port-group 1 的显示流量分担信息。

Switch# show port-group 1 load-balance
The loadbalance of the group 1 based on src MAC address.

3. 显示交换机的 port-group 1 的成员端口信息。

Switch# show port-group 1 port
　　　Selected
　　　Unselected

详细参数如表 1-25 所示。

表 1-25 端口信息表

显 示 内 容	解　　释
portnumber	端口号
port priority	端口优先级
system system	ID
system priority	System 优先级
LACP activety	端口是否以 active 模式加入组，如果是置 1
LACP timeout	端口超时方式，如果是短超时置 1
Aggregation	端口是否可汇聚，如果为 0 表示该端口为独立端口，不允许参与汇聚
Synchronization	端口是否与对端同步
Collecting	端口绑定状态机状态是否到达 collecting 状态
Distributing	端口绑定状态机状态是否到达 distributing 状态
Defaulted	本端端口是否使用默认的对端参数
Expired	端口收包状态机的状态是否处在 expire 状态
Selected	端口是否已被选择

4. 显示交换机的 port-group1 的汇聚端口信息。

Switch# show port-group 1 port-channel

```
Port channels in the group 1:
--------------------------------------------------------
Port-Channel: port-channel1
Number of port : 2          Standby port : NULL

Port in the port-channel :

Index                       Port              Mode
-------------------------------------------------------
1                           Ethernet0/0/1     active
2                           Ethernet0/0/2     active
```

详细参数如表 1-26 所示。

表 1-26　汇聚端口信息表

显示内容	解释
Port channels in the group	如果不存在 port-channel，则不会显示以上打印信息
Number of port	port-channel 中的端口数
Standby port	处于 standby 状态的端口，standby 的端口表示虽然端口满足进入 channel 的条件，但由于总共进入 channel 的端口数已经大于最大数，所以将该端口的状态设为 standby 而不是 selected

实训三　基于有 VLAN 的交换机链路聚合

一、实训目的

1. 使用两根网线做链路聚合，通过插拔线缆观察结果。
2. 把组作为交换机之间的 trunk 链路，实现跨交换机的 VLAN。

二、实训拓扑

该实训拓扑结构如图 1-38 所示。

三、实训要求

如图 1-38 所示进行连线。
交换机 A 和 B 做如下配置：
(1) 首先在 AB 交换机上启用生成树协议。
(2) 划分 Vlan10(9-15)、Vlan20(16-24)。

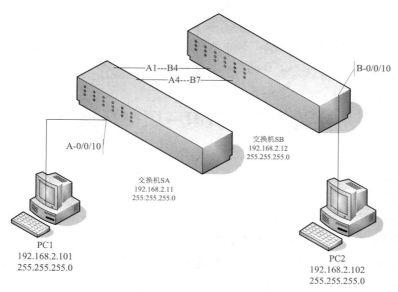

图 1-38 链路聚合实验拓扑图

(3) 划分 Trunk 端口 A-0/0/1、A-0/0/4；B-0/0/4、B-0/0/7。

(4) 做链路聚合 A-14 和 B-47，分别聚合生成聚合链路，此链路将作为两台交换机的 Trunk 链路。

实验结果：PC1---ping---PC2   通。

四、实训步骤

详细步骤说明：

(1) 选择对应设备，看图判断是否需要先做生成树；SA(config)#sp。

(2) 按要求连线。

4 条：   A1---B4   A4---B7   PC1---A 0/0/10   PC2---B 0/0/10。

(3) 配置相对应 IP 地址。

设备	IP 地址	Mask
交换机 A	192.168.2.11	255.255.255.0
交换机 B	192.168.2.12	255.255.255.0
A-PC1	192.168.2.101	255.255.255.0
B-PC2	192.168.2.102	255.255.255.0

(4) 并测试连通状态。

    A-PC1   ping 192.168.2.11 -t   ping 192.168.2.12 -t   ping 192.168.2.102 -t

    B-PC2   ping 192.168.2.12 -t   ping 192.168.2.11 -t   ping 192.168.2.101 -t

(5) 按要求配置，划分 VLAN 及端口并进行验证，观察 ping 窗口的连通变化。

    Vlan 10 (5-8) Vlan 20 (9-12) Vlan 30 (13-16)

(6) 交换机级联端口 A1、A4 和 B4、B7 设置为 Trunk，并进行验证，并观察 ping 窗口的连通变化。

(7) 聚合 A1、A4 和 B4、B7，新端口形成，并进行验证，观察 ping 窗口的连通变化。

(8) 观察 PC1---ping---PC2 通，实验成功。

(9) 按照作业内容，提取 A、B 两台设备配置，撰写实验报告。

实验十二　认识三层交换机

一、实验目的

1. 熟悉高端三层交换机的外观。
2. 了解高端三层交换机各端口的名称和作用。
3. 学会使用 TFTP 服务器对三层交换机版本进行升级和老版本的备份。
4. 学会使用相关命令对三层交换机的接口地址进行配置。

二、应用环境

交换机的分类方法有很多种，按照不同的原则，交换机可以分成各种不同的类别。按照网络 OSI 七层模型来划分，可以将交换机划分为二层交换机、三层交换机、多层交换机。二层交换机是按照 MAC 地址进行数据帧的过滤和转发，这种交换机是目前最常见的交换机。三层交换机采用"一次路由，多次交换"的原理，基于 IP 地址转发数据包。部分三层交换机也具有四层交换机的一些功能，譬如依据端口号进行转发。四层交换机以及四层以上的交换机都可以称为内容型交换机，一般使用在大型的网络数据中心。按照外观和架构的特点，可以将局域网交换机划分为机箱式交换机、机架式交换机、桌面式交换机。机箱式交换机外观比较庞大，这种交换机所有的部件都是可插拔的部件(一般称之为模块)，灵活性非常好。在实际的组网中，可以根据网络的要求选择不同的模块。模块可以分为几大类：一类是管理模块，它相当于计算机的主板和 CPU，用于管理整个交换机的工作；一类是应用模块，相当于计算机的 I/O 模块，负责连接其他网络设备和网络终端；另外还有电源模块、风扇模块等。在购买机箱式交换机时，需要分别购买机箱、管理模块、应用模块以及电源模块。机箱式交换机一般都是三层交换机或者多层交换机，在网络设计中，由于机箱式交换机性能和稳定性都比较卓越，因此价格比较昂贵，一般定位在核心层交换机或者汇聚层交换机。

三、实验设备

1. DCRS-7604(或 6804)交换机　　1 台。
2. PC 机　　1 台。
3. 交换机 console 线　　1 根。

四、实验拓扑

将 PC 机的串口和交换机的 console 口用 console 线，连接如图 1-39 所示。

图 1-39　实验拓扑图

五、实验要求

1. 正确认识交换机上各模块和物理端口名称。
2. 熟练对应物理端口在配置界面中对应的名称。

六、实验步骤

第一步：如图 1-40 所示，认识交换机的端口。

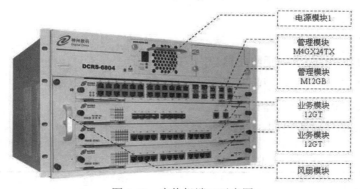

图 1-40　交换机端口示意图

第二步：以 MRS-7604-M12GB 为例，如图 1-41 所示，了解交换机的模块 MRS-7604-M12GB 是 DCRS-7604 交换机的主控交换模块，承担着系统状态的控制、路由的管理、用户接入的控制和管理、网络的维护。主控板插在单板插框的第 1、2 槽位中，支持主备冗余，实现热备份。同时它还带有 12 个 SFP 形式的千兆光纤接口。

交换机前面板 MRS-7604-M12GB 提供了 12 个 SFP 端口，同时还提供了 1 个 Console 配置口(即控制台)，以及 1 个 10/100Base-T Ethernet 口(即管理网口)。

图 1-41　MRS-7604-M12GB 前面板示意图

1. MRS-7604-M12GB 的前面板指示灯如表 1-27 所示。

表 1-27　MRS-7604-M12GB 指示灯说明

LED 指示灯	面板标记	状　态	含　义
电源指示灯	PWR	亮(绿色)	板卡加电
		灭	板卡断电
运行指示灯	RUN	亮(绿色 1Hz 闪烁)	板卡运行状态正常
		亮(绿色 8Hz 闪烁)	系统加载中
		亮(黄色 8Hz 闪烁)	系统关闭中
		亮(红色 8Hz 闪烁)	运行状态故障
		灭	板卡已关闭，可拔出
主备指示灯	M/S	亮(绿色)	主用
		灭	备用
风扇指示灯	FAN	亮(绿色)	风扇在位
		灭	风扇不在位
SFP 接口指示灯			
状态指示灯	Link	亮(绿色)	SFP 收发器网络连接正常
		灭	SFP 收发器没有网络连接
传输指示灯	Act	闪(绿)	正在发送或接收数据

MRS-7604-M12GB 提供了 12 个 SFP(Mini GBIC)千兆光纤收发器插槽。

MRS-7604-M12GB 支持以下类型的 SFP 收发器：

(1) SFP-SX 收发器。

(2) SFP-LX 10 公里收发器。

(3) SFP-LH-40 40 公里中距收发器。

(4) SFP-LH-70 70 公里长距收发器。

(5) SFP-LH-120 120 公里超长距收发器。

各 SFP 收发器传输距离如表 1-28 所示。

表 1-28　MRS-7604-M12GB 接口说明

接　口　形　式	规　　格
SFP-SX 收发器	62.5/125um 多模光纤：275m
	50/125um 多模光纤：550m
SFP-LX 收发器	9/125um 单模光纤：10km
SFP-LH-40 收发器	9/125um 单模光纤：40km
SFP-LH-70 收发器	9/125um 单模光纤：70km
SFP-LH-120 收发器	9/125um 单模光纤：120km

2. 前面板控制台。

MRS-7604-M12GB 提供了一个 RJ-45(母)Console 串口，如表 1-29 所示，通过该控制台，用户可以用来连接后台终端计算机，以进行系统的调试、配置、维护、管理、主机软件程序加载等工作。

表 1-29 MRS-7604-M12GB 控制台说明

属　性	规　　格
接头	RJ-45(母)
接口标准	RS-232
波特率	9600bps(默认)
支持服务	与字符终端相连
	与 PC 串口相连并在 PC 上运行终端仿真程序

3．前面板管理网口。

MRS-7604-M12GB 提供了一个 RJ-45(母)Ethernet 口，如表 1-30 所示，通过该网口，用户可以连接后台计算机，进行程序加载工作；也可通过该口接远端的网管工作站等设备，实现远程管理。

表 1-30 MRS-7604-M12GB 管理网口说明

属　性	规　　格
接头	RJ-45(母)
接口标准	10/100Mbps 自适应
	5 类非屏蔽双绞线(UTP)：300m

4．前面板复位按键。

MRS-7604-M12GB 提供了一个 RESET 复位按键，用于复位单板。MRS-7604-M12GB 提供了一个 SWAP 热拔按键，用于在设备运行过程中热拔此模块。当用户准备将模块拔下时应先按下 SWAP 按键。此时模块做热拔的准备工作，同时将系统运行指示灯(RUN)置为黄色 8 Hz 闪烁。当 RUN 灯灭后，表明板卡已关闭，可热拔出。

第三步：进入交换机配置界面。

```
DCRS-7604#show run
Current configuration:
!
        hostname DCRS-7604
!
Vlan 1
      vlan 1
!
Interface Ethernet1/1
!
Interface Ethernet1/2
……
Interface Ethernet2/27
!
Interface Ethernet2/28
!
```

Interface Ethernet0
!
DCRS-7604#

第四步：了解高端三层交换机各端口的名称。
Interface Ethernet1/1   表示第一模块的第 1 个接口；
Interface Ethernet1/28   表示第一模块的第 28 个接口；
Interface Ethernet2/1   表示第二模块的第 1 个接口；
Interface Ethernet2/28   表示第二模块的第 28 个接口；
Interface Ethernet0   表示 1 个 10/100Base-T Ethernet 口，即管理网口。

七、注意事项和排错

DCRS-7604/6804 等调试界面类似于 DCS-3926S，基本的命令格式都一样。

八、相关配置命令详解

clock set

命令：clock set <HH:MM:SS> <YYYY.MM.DD>。

功能：设置系统日期和时钟。

参数：<HH:MM:SS>为当前时钟，HH 取值范围为 0~23，MM 和 SS 取值范围为 0~59；<YYYY.MM.DD>为当前年、月和日，YYYY 取值范围为 1970~2100，MM 取值范围为 1~12，DD 取值范围为 1~31。

命令模式：特权用户配置模式。

默认情况：系统启动时默认为 2001 年 1 月 1 日 0：0：0。

使用指南：交换机在断电后不能继续计时，因此在要求使用确切时间的应用环境中，必须先设定交换机当前的日期和时间。

举例：设置交换机当前日期为 2002 年 8 月 1 日 23 时 0 分 0 秒。

SWITCH#clock set 23:0:0 2002.8.1
相关命令：show clock

config

命令：config [terminal]。

功能：从特权用户配置模式进入到全局配置模式。

参数：[terminal]表示进行终端配置。

命令模式：特权用户配置模式。

举例：SWITCH#config。

enable

命令：enable。

功能：用户使用 enable 命令，从普通用户配置模式进入特权用户配置模式。

命令模式：一般用户配置模式。

使用指南：为了防止非特权用户的非法访问，在从一般用户配置模式进入到特权用户配置模式时，需要进行用户安全验证，即需要输入特权用户密码。当正确地输入特权用户密码后，将进入特权用户配置模式，如果连续输入 3 次特权用户密码均不正确，则保持当前的一般用户配置模式不变。通过使用全局配置模式下的"enable password"命令来设置特权用户密码。

举例：

SWITCH>enable
password：*****(admin)
SWITCH#

相关命令：enable password。

enable password

命令：enable password。

功能：修改从普通用户配置模式进入特权用户配置模式的口令，输入本命令后直接回车将出现<Current password>、<New password>参数，需要用户配置。

参数：<Current password>为原来的密码，最长不超过 16 个字符；<New password>为新的密码，最长不超过 16 个字符；<Confirm new password>为确认新密码，值应与新密码一样，否则需要重新设置密码。

命令模式：全局配置模式。

默认情况：系统默认的特权用户口令为空。当用户是首次配置时，出现要输入原密码的提示时，直接回车即可。

使用指南：配置特权用户口令，可以防止非特权用户的非法侵入，建议网络管理员在首次配置交换机时就设定特权用户口令。另外当管理员需要长时间离开终端屏幕时，最好执行 exit 命令退出特权用户配置模式。

举例：设置特权用户的口令为 admin。

SWITCH(Config)#enable password
Current password: (首次配置，没有设置任何口令，直接回车)
New password:*****   (设置新的口令为 admin)
Confirm new password:*****(确认新的口令 admin)
SWITCH(Config)#

相关命令：enable。

exec timeout

命令：exec timeout <minutes>。

功能：设置退出特权用户配置模式超时时间。

参数：<minute>为时间值，单位为分钟，取值范围为0~300。

命令模式：全局配置模式。

默认情况：系统默认为5分钟。

使用指南：为确保交换机使用的安全性，防止非法用户的恶意操作，当特权用户在做完最后一项配置后，开始计时，到达设置时间值时，系统就自动退出特权用户配置模式。用户如果想再次进入特权用户配置模式，需要再次输入特权用户密码和口令。如果配置 exec timeout 的值为0，则说明不会退出特权用户配置模式。

举例：设置交换机退出特权用户配置模式的超时时间为6分钟。

SWITCH(Config)#exec timeout 6

exit

命令：exit。

功能：从当前模式退出，进入上一个模式，如在全局配置模式使用本命令退回到特权用户配置模式，在特权用户配置模式使用本命令退回到一般用户配置模式等。

命令模式：各种配置模式。

举例：

SWITCH#exit
SWITCH>

help

命令：help。

功能：输出有关命令解释器帮助系统的简单描述。

命令模式：各种配置模式。

使用指南：交换机提供随时随地的在线帮助，help 命令则显示关于整个帮助体系的信息，包括完全帮助和部分帮助，用户可以随时随地通过键盘输入？获取在线帮助。

举例：

SWITCH>help
enable   -- Enable Privileged mode
exit   -- Exit telnet session
help   -- help
show   -- Show running system information

ip host

命令：ip host <hostname> <ip_addr>。

　　　　no ip host <hostname>。

功能：设置主机与 IP 地址映射关系；本命令的 no 操作为删除该项映射关系。

参数：<hostname>为主机名称，最长不超过15个字符；<ip_addr>为主机名相应 IP 地址，点分十进制格式。

命令模式：全局配置模式。

使用指南：设置一个确定的主机和 IP 地址的对应关系，可用于如"ping <host>"等命令中。

举例：设置主机名为 beijing 的主机的 IP 地址为 200.121.1.1。

SWITCH(Config)#ip host beijing 200.121.1.1

相关命令：telnet、ping、traceroute。

hostname

命令：hostname <hostname>。
功能：设置交换机命令行界面的提示符。
参数：<hostname>为提示符的字符串，最长不超过 30 个字符。
命令模式：全局配置模式。
默认情况：系统默认提示符为"DCRS-7604"。
使用指南：通过本命令用户可以根据实际情况设置交换机命令行的提示符。
举例：设置提示符为 Test。

SWITCH(Config)#hostname Test
Test(Config)#

reload

命令：reload。
功能：热启动交换机。
命令模式：特权用户配置模式。
使用指南：用户可以通过本命令，在不关闭电源的情况下，重新启动交换机。

set default

命令：set default。
功能：恢复交换机的出厂设置。
命令模式：特权用户配置模式。
使用指南：恢复交换机的出厂设置，即用户对交换机做的所有配置都消失，用户重新启动交换机后，出现的提示与交换机首次上电一样。
注意：配置本命令后，必须执行write命令，进行配置保留后重启交换机即可使交换机恢复到出厂设置。
举例：

SWITCH#set default
Are you sure? [Y/N] = y
SWITCH#write
SWITCH#reload

setup

命令：setup。

功能：进入交换机的 Setup 配置模式。

命令模式：特权用户配置模式。

使用指南：DCRS-7604 提供 Setup 配置模式，在 Setup 配置模式下用户可进行 IP 地址等的配置。

language

命令：language {chinese|english}。

功能：设置显示的帮助信息的语言类型。

参数：chinese 为中文显示；english 为英文显示。

命令模式：特权用户配置模式。

默认情况：系统默认英文显示。

使用指南：DCRS-7604 提供了两种语言的帮助信息，用户可根据自己的喜好选择语言类型。系统若重启后，帮助显示信息恢复为英文显示。

write

命令：write。

功能：将当前运行时配置参数保存到 Flash Memory。

命令模式：特权用户配置模式。

使用指南：当完成一组配置，并且已经达到预定功能，应将当前配置保存到 Flash 中，以便因不慎关机或断电时，系统可以自动恢复到原先保存的配置。相当于 copy running-config startup-config 命令。

相关命令：copy running-config startup-config。

九、课后练习

1. 熟悉交换机的各种配置模式。
2. 熟悉交换机的 CLI 界面调试技巧。
3. 交换机恢复出厂设置及其基本配置。
4. 使用 telnet 方式管理交换机。
5. 使用 Web 方式管理交换机。
6. 交换机文件备份。
7. 交换机系统升级和文件还原。
8. 密码丢失的解决方法。
9. Bootrom 下的升级配置。

实验十三　多层交换机 VLAN 的划分和 VLAN 间路由

一、实验目的

1. 了解 VLAN 原理。
2. 学会使用各种多层交换设备进行 VLAN 的划分。
3. 理解 VLAN 之间路由的原理和实现方法。

二、应用环境

软件实验室的 IP 地址段是 192.168.10.0/24，多媒体实验室的 IP 地址段是 192.168.20.0/24，为了保证它们之间的数据互不干扰，也不影响各自的通信效率，我们划分了 VLAN，使两个实验室属于不同的 VLAN。两个实验室有时候也需要相互通信，此时就要利用三层交换机划分 VLAN。

三、实验设备

1. DCRS-7604(或 6804 或 5526S)交换机　　1 台。
2. PC 机　　2 台。
3. Console 线　　1 根。
4. 直通网线若干。

四、实验拓扑

使用一台交换机和两台 PC 机，将其中 PC2 作为控制台终端，使用 Console 口配置方式；使用两根网线分别将 PC1 和 PC2 连接到交换机的 RJ-45 接口上。该实验拓扑结构如图 1-42 所示。

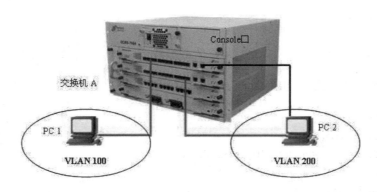

图 1-42　三层交换机拓扑图

五、实验要求

在交换机上划分两个基于端口的 VLAN：VLAN100、VLAN200。使得 VLAN100 的成员能够互相访问，VLAN200 的成员能够互相访问；VLAN100 和 VLAN200 成员之间不能互相访问。如表 1-31 所示。

表 1-31 VLAN 划分表

VLAN	端口成员
100	1/1~1/12
200	1/13~1/24

PC1 和 PC2 的网络设置如表 1-32 所示。

表 1-32 PC1/PC2 配置表

设备	端口	IP1	网关 1	IP2	网关 2	Mask
交换机 A		192.168.2.1	无	192.168.2.1	无	255.255.255.0
Vlan 100		无	无	192.168.10.1	无	255.255.255.0
Vlan 200		无	无	192.168.20.1	无	255.255.255.0
PC1	1~12	192.168.2.101	无	192.168.10.101	192.168.10.1	255.255.255.0
PC2	13~24	192.168.2.102	无	192.168.20.102	192.168.20.1	255.255.255.0

各设备的 IP 地址首先按照 IP1 配置，使用 pc1 ping pc2，应该不通；再按照 IP2 配置地址，并在交换机上配置 VLAN 接口 IP 地址，使用 pc1 ping pc2，则通，该通信属于 VLAN 间通信，要经过三层设备的路由。

若实验结果和理论相符，则本实验完成。

六、实验步骤

第一步：交换机恢复出厂设置。

switch#set default
switch#write
switch#reload

第二步：给交换机设置 IP 地址即管理 IP。

switch#config
switch(Config)#interface vlan 1
switch(Config-If-Vlan1)#ip address 192.168.2.1 255.255.255.0
switch(Config-If-Vlan1)#no shutdown
switch(Config-If-Vlan1)#exit
switch(Config)#exit

第三步：创建 VLAN100 和 VLAN200。

switch(Config)#
switch(Config)#vlan 100
switch(Config-Vlan100)#exit
switch(Config)#vlan 200
switch(Config-Vlan200)#exit
switch(Config)#

验证配置：

switch#show vlan

VLAN	Name	Type	Media	Ports	
1	default	Static	ENET	Ethernet1/1	Ethernet1/2
				Ethernet1/3	Ethernet1/4
				Ethernet1/5	Ethernet1/6
				Ethernet1/7	Ethernet1/8
				Ethernet1/9	Ethernet1/10
				Ethernet1/11	Ethernet1/12
				Ethernet1/13	Ethernet1/14
				Ethernet1/15	Ethernet1/16
				Ethernet1/17	Ethernet1/18
				Ethernet1/19	Ethernet1/20
				Ethernet1/21	Ethernet1/22
				Ethernet1/23	Ethernet1/24
				Ethernet1/25	Ethernet1/26
				Ethernet1/27	Ethernet1/28
100	VLAN0100	Static	ENET		
200	VLAN0200	Static	ENET		

第四步：给 VLAN100 和 VLAN200 添加端口。

switch(Config)#vlan 100　　　　　　！进入 vlan 100
switch(Config-Vlan100)#switchport interface ethernet 1/1-12
Set the port Ethernet1/1 access vlan 100 successfully
Set the port Ethernet1/2 access vlan 100 successfully
Set the port Ethernet1/3 access vlan 100 successfully
Set the port Ethernet1/4 access vlan 100 successfully
Set the port Ethernet1/5 access vlan 100 successfully
Set the port Ethernet1/6 access vlan 100 successfully
Set the port Ethernet1/7 access vlan 100 successfully
Set the port Ethernet1/8 access vlan 100 successfully
Set the port Ethernet1/9 access vlan 100 successfully
Set the port Ethernet1/10 access vlan 100 successfully
Set the port Ethernet1/11 access vlan 100 successfully

```
Set the port Ethernet1/12 access vlan 100 successfully
switch(Config-Vlan100)#exit
switch(Config)#vlan 200                    ！进入 vlan 200
switch(Config-Vlan200)#switchport interface ethernet 1/13-24
Set the port Ethernet1/13 access vlan 200 successfully
Set the port Ethernet1/14 access vlan 200 successfully
Set the port Ethernet1/15 access vlan 200 successfully
Set the port Ethernet1/16 access vlan 200 successfully
Set the port Ethernet1/17 access vlan 200 successfully
Set the port Ethernet1/18 access vlan 200 successfully
Set the port Ethernet1/19 access vlan 200 successfully
Set the port Ethernet1/20 access vlan 200 successfully
Set the port Ethernet1/21 access vlan 200 successfully
Set the port Ethernet1/22 access vlan 200 successfully
Set the port Ethernet1/23 access vlan 200 successfully
Set the port Ethernet1/24 access vlan 200 successfully
switch(Config-Vlan200)#exit
```

验证配置：

```
switch#show vlan
VLAN Name           Type          Media           Ports
---- --------------- ----------------- ---------------  ---------------------------------------
1    default        Static ENET    Ethernet1/25    Ethernet1/26
                                   Ethernet1/27    Ethernet1/28
100  VLAN0100       Static ENET    Ethernet1/1     Ethernet1/2
                                   Ethernet1/3     Ethernet1/4
                                   Ethernet1/5     Ethernet1/6
                                   Ethernet1/7     Ethernet1/8
                                   Ethernet1/9     Ethernet1/10
                                   Ethernet1/11    Ethernet1/12
200  VLAN0200       Static ENET    Ethernet1/13    Ethernet1/14
                                   Ethernet1/15    Ethernet1/16
                                   Ethernet1/17    Ethernet1/18
                                   Ethernet1/19    Ethernet1/20
                                   Ethernet1/21    Ethernet1/22
                                   Ethernet1/23    Ethernet1/24
switch#
```

第五步：验证实验。结果如表 1-33 所示。

表 1-33　验证结果

PC1 的位置	PC2 的位置	动作	结果
1/1-1/12 端口		1/13-1/24 端口 PC1 ping	不通

第六步：添加 VLAN 地址。

switch(Config)#interface vlan 100
switch(Config-If-Vlan100)#%Feb 13　15:47:45 2006 %LINK-5-CHANGED:　Interface Vlan100, changed state to UP%Feb 13 15:47:45 2006 %LINEPROTO-5-UPDOWN: Line protocol
on Interface Vlan100, changed state to UP
switch(Config-If-Vlan100)#
switch(Config-If-Vlan100)#ip address 192.168.10.1 255.255.255.0
switch(Config-If-Vlan100)#no shut
switch(Config-If-Vlan100)#exit
switch(Config)#interface vlan 200
switch(Config-If-Vlan200)#%Feb 13　15:48:06 2006 %LINK-5-CHANGED:　Interface Vlan200, changed　state to UPswitch(Config-If-Vlan200)#ip address 192.168.20.1 255.255.255.0
switch(Config-If-Vlan200)#no shut
switch(Config-If-Vlan200)#exit
switch(Config)#

验证配置：

switch#show ip route
Total route items is 3, the matched route items is 3
Codes: C - connected, S - static, R - RIP derived, O - OSPF derived
　　　A - OSPF ASE, B - BGP derived, D - DVMRP derived

	Destination	Mask	Nexthop	Interface	Preference
C	192.168.2.0	255.255.255.0	0.0.0.0	Vlan1	0
C	192.168.10.0	255.255.255.0	0.0.0.0	Vlan100	0
C	192.168.20.0	255.255.255.0	0.0.0.0	Vlan200	0

switch#

第七步：验证实验。结果如表 1-34 所示。

表 1-34　验证结果

PC1 的位置	PC2 的位置	动作	结果
1/1-1/12 端口	1/13-1/24 端口	PC1 ping PC2	通

七、注意事项和排错

三层交换机可以在多个 VLAN 接口上配置 IP 地址。

八、配置序列

switch#show run
Current configuration:

```
!
    hostname switch
!
Vlan 1
    vlan 1
!
Vlan 100
    vlan 100
!
Vlan 200
    vlan 200
!
Interface Ethernet1/1
    switchport access vlan 100
!
Interface Ethernet1/2
    switchport access vlan 100
!
Interface Ethernet1/3
    switchport access vlan 100
!
Interface Ethernet1/4
    switchport access vlan 100
!
Interface Ethernet1/5
    switchport access vlan 100
!
Interface Ethernet1/6
    switchport access vlan 100
!
Interface Ethernet1/7
    switchport access vlan 100
!
Interface Ethernet1/8
    switchport access vlan 100
!
Interface Ethernet1/9
    switchport access vlan 100
!
Interface Ethernet1/10
    switchport access vlan 100
!
Interface Ethernet1/11
    switchport access vlan 100
```

!
Interface Ethernet1/12
    switchport access vlan 100
!
Interface Ethernet1/13
    switchport access vlan 200
!
Interface Ethernet1/14
    switchport access vlan 200
!
Interface Ethernet1/15
    switchport access vlan 200
!
Interface Ethernet1/16
    switchport access vlan 200
!
Interface Ethernet1/17
    switchport access vlan 200
!
Interface Ethernet1/18
    switchport access vlan 200
!
Interface Ethernet1/19
    switchport access vlan 200
!
Interface Ethernet1/20
    switchport access vlan 200
!
Interface Ethernet1/21
    switchport access vlan 200
!
Interface Ethernet1/22
    switchport access vlan 200
!
Interface Ethernet1/23
    switchport access vlan 200
!
Interface Ethernet1/24
    switchport access vlan 200
!
Interface Ethernet1/25
!
Interface Ethernet1/26
!

```
Interface Ethernet1/27
!
Interface Ethernet1/28
!
!
interface Vlan1
    interface vlan 1
    ip address 192.168.2.1 255.255.255.0
!
interface Vlan100
    interface vlan 100
    ip address 192.168.10.1 255.255.255.0
!
interface Vlan200
    interface vlan 200
    ip address 192.168.20.1 255.255.255.0
!
Interface Ethernet0
switch#
```

九、思考题

如果第二次配置 IP 地址时，没有给 PC 机配置网关，请问还会通信么？为什么？

十、课后练习

请给交换机划分多个 VLAN，验证 VLAN 实验。

十一、相关配置命令详解

vlan

命令：vlan <vlan-id>。

　　　　no vlan <vlan-id>。

功能：创建 VLAN 并且进入 VLAN 配置模式，在 VLAN 的模式中，用户可以配置 VLAN 名称和为该 VLAN 分配交换机端口；本命令的 no 操作为删除指定的 VLAN。

参数：<vlan-id>为要创建/删除的 VLAN 的 VID，取值范围为 1~4094。

命令模式：全局配置模式。

默认情况：交换机默认只有 VLAN1。

使用指南：VLAN1 为交换机的默认 VLAN，用户不能配置和删除 VLAN1。允许配置 VLAN 的总共数量为 4094 个。另需要注意的是不能使用本命令删除通过 GVRP 学习到的动态 VLAN。

举例：创建 VLAN100，并且进入 VLAN100 的配置模式。

DCRS-7604 (Config)#vlan 100

DCRS-7604 (Config-Vlan100)#

name

命令：name <vlan-name>。

　　　　no name。

功能：为 VLAN 指定名称，VLAN 的名称是对该 VLAN 的一个描述性字符串；本命令的 no 操作为删除 VLAN 的名称。

参数：<vlan-name>为指定的 VLAN 名称字符串。

命令模式：VLAN 配置模式。

默认情况：VLAN 默认名称为 vlanXXX，其中 XXX 为 VID。

使用指南：交换机提供为不同的 VLAN 指定名称的功能，有助于用户记忆 VLAN，方便管理。

举例：为 VLAN100 指定名称为 TestVlan。

DCRS-7604 (Config-Vlan100)#name TestVlan

switchport access vlan

命令：switchport access vlan <vlan-id>。

　　　　no switchport access vlan。

功能：将当前 Access 端口加入到指定 VLAN；本命令 no 操作为将当前端口从 VLAN 里删除。

参数：<vlan-id>为当前端口要加入的 vlanVID，取值范围为 1~4094。

命令模式：接口配置模式。

默认情况：所有端口默认属于 VLAN1。

使用指南：只有属于 Access mode 的端口才能加入到指定的 VLAN 中，并且 Access 端口同时只能加入到一个 VLAN 里去。

举例：设置某 Access 端口加入 VLAN100。

DCRS-7604 (Config)#interface ethernet 1/8

DCRS-7604 (Config-ethernet1/8)#switchport mode access

DCRS-7604 (Config-ethernet1/8)#switchport access vlan 100

DCRS-7604 (Config-ethernet1/8)#exit

switchport interface

命令：switchport interface <interface-list>。

　　　　no switchport interface <interface-list>。

功能：给 VLAN 分配以太网端口的命令；本命令的 no 操作为删除指定 VLAN 内的一个或一组端口。

参数：<interface-list>为要添加或者删除的端口的列表，支持";"、"-"，如：ethernet 1/1;2;5 或 ethernet 1/1-6;8。

命令模式：VLAN 配置模式。

默认情况：新建立的 VLAN 默认不包含任何端口。

使用指南：Access 端口为普通端口，可以加入 VLAN，但同时只允许加入一个 VLAN。

举例：为 VLAN100 分配百兆以太网端口 1、3、4~7、8。

DCRS-7604 (Config-Vlan100)#switchport interface ethernet 1/1;3;4-7;8

switchport mode

命令：switchport mode {trunk|access}。

功能：设置交换机的端口为 access 模式或者 trunk 模式。

参数：trunk 表示端口允许通过多个 VLAN 流量；access 为端口只能属于一个 VLAN。

命令模式：接口配置模式。

默认情况：端口默认为 Access 模式。

使用指南：工作在 trunk mode 下的端口称为 Trunk 端口，Trunk 端口可以通过多个 VLAN 的流量，通过 Trunk 端口之间的互联，可以实现不同交换机上的相同 VLAN 的互通；工作在 access mode 下的端口称为 Access 端口，Access 端口可以分配给一个 VLAN，并且同时只能分配给一个 VLAN。

举例：将端口 5 设置为 trunk 模式，端口 8 设置为 access 模式。

DCRS-7604 (Config)#interface ethernet 1/5
DCRS-7604 (Config-ethernet1/5)#switchport mode trunk
DCRS-7604 (Config-ethernet1/5)#exit
DCRS-7604 (Config)#interface ethernet 1/8
DCRS-7604 (Config-ethernet1/8)#switchport mode access
DCRS-7604 (Config-ethernet1/8)#exit

switchport trunk allowed vlan

命令：switchport trunk allowed vlan {<vlan-list>|all}。

　　　　no switchport trunk allowed vlan。

功能：设置 Trunk 端口允许通过 VLAN；本命令的 no 操作为恢复默认情况。

参数：<vlan-list>为允许在该 Trunk 端口上通过的 VLAN 列表；all 关键字表示允许该 Trunk 端口通过所有 VLAN 的流量。

命令模式：接口配置模式。

默认情况：Trunk 端口默认允许通过所有 VLAN。

使用指南：用户可以通过本命令设置哪些 VLAN 的流量通过 Trunk 端口，没有包含的 VLAN 流量则被禁止。

举例：设置 Trunk 端口允许通过 VLAN1、3、5~20 的流量。

DCRS-7604 (Config)#interface ethernet 1/5
DCRS-7604 (Config-ethernet1/5)#switchport mode trunk

DCRS-7604 (Config-ethernet1/5)#switchport trunk allowed vlan 1;3;5-20
DCRS-7604 (Config-ethernet1/5)#exit

switchport trunk native vlan

命令：switchport trunk native vlan <vlan-id>。
　　　no switchport trunk native vlan。
功能：设置 Trunk 端口的 PVID；本命令的 no 操作为恢复默认值。
参数：<vlan-id>为 Trunk 端口的 PVID。
命令模式：接口配置模式。
默认情况：Trunk 端口默认的 PVID 为 1。
使用指南：在 802.1Q中定义了PVID这个概念。Trunk端口的PVID的作用是当一个untagged的帧进入Trunk端口，端口会对这个untagged帧打上带有本命令设置的native PVID的tag标记，用于VLAN的转发。
举例：设置某 Trunk 端口的 native vlan 为 100。

DCRS-7604 (Config)#interface ethernet 1/5
DCRS-7604 (Config-ethernet1/5)#switchport mode trunk
DCRS-7604 (Config-ethernet1/5)#switchport trunk native vlan 100
DCRS-7604 (Config-ethernet1/5)#exit

vlan ingress disable

命令：vlan ingress disable。
　　　no vlan ingress disable。
功能：关闭端口的 VLAN 入口规则；本命令的 no 操作为打开入口准则。
命令模式：接口配置模式。
默认情况：系统默认打开端口的 VLAN 入口准则。
使用指南：当打开端口的 VLAN 入口规则，系统在接收数据时会检查源端口是否是该 VLAN 的成员端口，如果是则接收数据并转发到目的端口，否则丢弃该数据。
举例：关闭端口的 VLAN 入口规则。

DCRS-7604 (Config-Ethernet1/1)# vlan ingress disable

实验十四　多层交换机实现二层交换机 VLAN 之间路由

一、实验目的

1. 理解多层交换机的路由原理。
2. 了解多层交换机在实际网络中的常用配置。

3. 回顾二层交换机 VLAN 的划分方法。
4. 进一步理解 802.1Q 的原理和使用方法。

二、应用环境

二层交换机属于接入层交换机，在二层交换机上根据连接用户的不同，划分了不同 VLAN，有时候会出现同一个 VLAN 处于不同的交换机上。这些二层交换机被一台三层交换机所汇聚。因此我们需要实现多交换机的跨交换机 VLAN 通信，也需要实现 VLAN 间的通信。因此出现本实验所要演示的功能。

三、实验设备

1. DCRS-7604(或 6804 或 5526S)交换机 1 台。
2. DCS-3926S 交换机 1~2 台。
3. PC 机 2~4 台。
4. Console 线 1~3 根。
5. 直通网线若干。

四、实验拓扑

多层交换机实现二层交换机 VLAN 之间路由的实验拓扑图如图 1-43 所示。

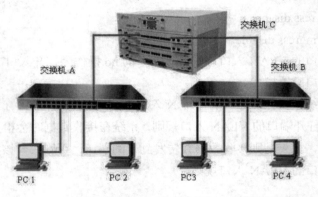

图 1-43 实验拓扑图

五、实验要求

在交换机 A 和交换机 B 上分别划分两个基于端口的 VLAN：VLAN100、VLAN200。如表 1-35 所示。

表 1-35 交换机 A/B 的 VLAN 划分

VLAN	端口成员
100	0/0/1~0/0/8
200	0/0/9~0/0/16
Trunk 口	24

在交换机 C 上也划分两个基于端口的 VLAN：VLAN100、VLAN200。把端口 1 和端口 2 都设置成 Trunk 口。如表 1-36 所示。

表 1-36　VLAN 的配置

VLAN	IP	Mask
100	192.168.10.1	255.255.255.0
200	192.168.20.1	255.255.255.0
Trunk 口		0/0/1 和 0/0/2

交换机 A 的 24 口连接交换机 C 的 1 口，交换机 B 的 24 口连接交换机 C 的 2 口。如表 1-37 所示。

表 1-37　PC 的配置

设备	IP 地址	gateway	Mask
PC1	192.168.10.11	192.168.10.1	255.255.255.0
PC2	192.168.20.22	192.168.20.1	255.255.255.0
PC3	192.168.10.33	192.168.10.1	255.255.255.0

验证：

1. 不给 PC 设置网关：PC1、PC3 分别接在不同交换机 VLAN100 的成员端口 1~8 上，两台 PC 互相可以 ping 通；PC2、PC4 分别接在不同交换机 VLAN 的成员端口 9~16 上，两台 PC 互相可以 ping 通；PC1、PC3 和 PC2、PC4 接在不同 VLAN 的成员端口上则互相 ping 不通。

2. 给 PC 设置网关：PC1、PC3 和 PC2、PC4 接在不同 VLAN 的成员端口上也可以互相 ping 通。

若实验结果和理论相符，则本实验完成。

六、实验步骤

第一步：交换机恢复出厂设置。

switch#set default
switch#write
switch#reload

第二步：给交换机设置标示符和管理 IP。
交换机 A：

switch(Config)#hostname switchA
switchA(Config)#interface vlan 1
switchA(Config-If-Vlan1)#ip address 192.168.2.11 255.255.255.0
switchA(Config-If-Vlan1)#no shutdown
switchA(Config-If-Vlan1)#exit

交换机 B：

switch(Config)#hostname switchB
switchB(Config)#interface vlan 1
switchB(Config-If-Vlan1)#ip address 192.168.2.12 255.255.255.0
switchB(Config-If-Vlan1)#no shutdown
switchB(Config-If-Vlan1)#exit
switchB(Config)#

交换机 C：

DCRS-7604#config
DCRS-7604(Config)#
DCRS-7604(Config)#hostname switchC
switchC(Config)#interface vlan 1
switchC(Config-If-Vlan1)#ip address 192.168.2.13 255.255.255.0
switchC(Config-If-Vlan1)#no shutdown
switchC(Config-If-Vlan1)#exit
switchC(Config)#exit
switchC#

第三步：在交换机中创建 VLAN100 和 VLAN200，并添加端口。

交换机 A：

switchA(Config)#vlan 100
switchA(Config-Vlan100)#
switchA(Config-Vlan100)#switchport interface ethernet 0/0/1-8
switchA(Config-Vlan100)#exit
switchA(Config)#vlan 200
switchA(Config-Vlan200)#switchport interface ethernet 0/0/9-16
switchA(Config-Vlan200)#exit
switchA(Config)#

验证配置：

switchA#show vlan

VLAN	Name	Type	Media	Ports	
1	default	Static	ENET	Ethernet0/0/17	Ethernet0/0/18
				Ethernet0/0/19	Ethernet0/0/20
				Ethernet0/0/21	Ethernet0/0/22
				Ethernet0/0/23	Ethernet0/0/24
100	VLAN0100	Static	ENET	Ethernet0/0/1	Ethernet0/0/2
				Ethernet0/0/3	Ethernet0/0/4
				Ethernet0/0/5	Ethernet0/0/6

200	VLAN0200	Static	ENET	Ethernet0/0/7 Ethernet0/0/9 Ethernet0/0/11 Ethernet0/0/13 Ethernet0/0/15	Ethernet0/0/8 Ethernet0/0/10 Ethernet0/0/12 Ethernet0/0/14 Ethernet0/0/16

switchA#

交换机 B：配置与交换机 A 一样。

第四步：设置交换机 trunk 端口。

交换机 A：

switchA(Config)#interface ethernet 0/0/24
switchA(Config-Ethernet0/0/24)#switchport mode trunk
Set the port Ethernet0/0/24 mode TRUNK successfully
switchA(Config-Ethernet0/0/24)#switchport trunk allowed vlan all
set the port Ethernet0/0/24 allowed vlan successfully
switchA(Config-Ethernet0/0/24)#exit
switchA(Config)#

验证配置：

switchA#show vlan

VLAN	Name	Type	Media	Ports	
1	default	Static	ENET	Ethernet0/0/17 Ethernet0/0/19 Ethernet0/0/21 Ethernet0/0/23 Ethernet0/0/24(T)	Ethernet0/0/18 Ethernet0/0/20 Ethernet0/0/22
100	VLAN0100	Static	ENET	Ethernet0/0/1 Ethernet0/0/3 Ethernet0/0/5 Ethernet0/0/7 Ethernet0/0/24(T)	Ethernet0/0/2 Ethernet0/0/4 Ethernet0/0/6 Ethernet0/0/8
200	VLAN0200	Static	ENET	Ethernet0/0/9 Ethernet0/0/11 Ethernet0/0/13 Ethernet0/0/15 Ethernet0/0/24(T)	Ethernet0/0/10 Ethernet0/0/12 Ethernet0/0/14 Ethernet0/0/16

switchA#

24 口已经出现在 VLAN1、VLAN100 和 VLAN200 中，并且 24 口不是一个普通端口，是 tagged 端口。

交换机 B：配置同交换机 A。

交换机 C：

```
switchC(Config)#vlan 100
switchC(Config-Vlan100)#exit
switchC(Config)#vlan 200
switchC(Config-Vlan200)#exit
switchC(Config)#interface ethernet 0/0/1-2
switchC(Config-Port-Range)#switchport mode trunk
Set the port Ethernet 0/0/1 mode TRUNK successfully
Set the port Ethernet 0/0/2 mode TRUNK successfully
switchC(Config-Port-Range)#switchport trunk allowed vlan all
set the port Ethernet 0/0/1 allowed vlan successfully
set the port Ethernet 0/0/2 allowed vlan successfully
switchC(Config-Port-Range)#exit
switchC(Config)#exit
```

验证配置：

```
switchC#show vlan
VLAN Name           Type      Media   Ports
1    default        Static    ENET    Ethernet1/1(T)    Ethernet1/2(T)
                                      Ethernet1/3       Ethernet1/4
                                      Ethernet1/5       Ethernet1/6
                                      Ethernet1/7       Ethernet1/8
                                      Ethernet1/9       Ethernet1/10
                                      Ethernet1/11      Ethernet1/12
                                      Ethernet1/13      Ethernet1/14
                                      Ethernet1/15      Ethernet1/16
                                      Ethernet1/17      Ethernet1/18
                                      Ethernet1/19      Ethernet1/20
                                      Ethernet1/21      Ethernet1/22
                                      Ethernet1/23      Ethernet1/24
                                      Ethernet1/25      Ethernet1/26
                                      Ethernet1/27      Ethernet1/28
100  VLAN0100       Static    ENET    Ethernet1/1(T)    Ethernet1/2(T)
200  VLAN0200       Static    ENET    Ethernet1/1(T)    Ethernet1/2(T)
switchC#
```

第五步：交换机 C 添加 VLAN 地址。

```
switchC(Config)#interface vlan 100
switchC(Config-If-Vlan100)#ip address 192.168.10.1 255.255.255.0
switchC(Config-If-Vlan100)#no shut
switchC(Config-If-Vlan100)#exit
switchC(Config)#interface vlan 200
switchC(Config-If-Vlan200)#ip address 192.168.20.1 255.255.255.0
switchC(Config-If-Vlan200)#no shutdown
switchC(Config-If-Vlan200)#exit
```

switchC(Config)#

验证配置：

switch#show ip route
Total route items is 3, the matched route items is 3
Codes: C - connected, S - static, R - RIP derived, O - OSPF derived
       A - OSPF ASE, B - BGP derived, D - DVMRP derived

Destination	Mask	Nexthop	Interface	Preference
C 192.168.2.0	255.255.255.0	0.0.0.0	Vlan1	0
C 192.168.10.0	255.255.255.0	0.0.0.0	Vlan100	0
C 192.168.20.0	255.255.255.0	0.0.0.0	Vlan200	0

switch#

第六步：验证实验。
1. PC 不配置网关，互相 ping，查看结果。
2. PC 配置网关，互相 ping，查看结果。

七、注意事项和排错

Show ip route 时，如果在某一个网段上没有 active 的设备连接在三层交换机上，则这个网段的路由不会被显示出来。

八、思考题

如果两台三层交换机级联，如何进行 VLAN 的配置？需要把某些端口的模式设置为 trunk 么？

九、相关配置命令详解

vlan

命令：vlan <vlan-id>。
　　　no vlan <vlan-id>。

功能：创建 VLAN 并且进入 VLAN 配置模式，在 VLAN 模式中，用户可以配置 VLAN 名称和为该 VLAN 分配交换机端口；本命令的 no 操作为删除指定的 VLAN。

参数：<vlan-id>为要创建/删除的 VLAN 的 VID，取值范围为 1~4094。

命令模式：全局配置模式。

默认情况：交换机默认只有 VLAN1。

使用指南：VLAN1 为交换机的默认 VLAN，用户不能配置和删除 VLAN1。允许配置 VLAN 的总共数量为 4094 个。另需要注意的是不能使用本命令删除通过 GVRP 学习到的动态 VLAN。

举例：创建 VLAN100，并且进入 VLAN100 的配置模式。

DCRS-7604 (Config)#vlan 100
DCRS-7604 (Config-Vlan100)#

name

命令：name <vlan-name>。

　　　no name。

功能：为 VLAN 指定名称，VLAN 的名称是对该 VLAN 的一个描述性字符串；本命令的 no 操作为删除 VLAN 的名称。

参数：<vlan-name>为指定的 VLAN 名称字符串。

命令模式：VLAN 配置模式。

默认情况：VLAN 默认名称为 vlanXXX，其中 XXX 为 VID。

使用指南：交换机提供为不同的 VLAN 指定名称的功能，有助于用户记忆 VLAN，方便管理。

举例：为 VLAN100 指定名称为 TestVlan。

DCRS-7604 (Config-Vlan100)#name TestVlan

switchport access vlan

命令：switchport access vlan <vlan-id>。

　　　no switchport access vlan。

功能：将当前 Access 端口加入到指定 VLAN；本命令 no 操作为将当前端口从 VLAN 里删除。

参数：<vlan-id>为当前端口要加入的 vlanVID，取值范围为 1~4094。

命令模式：接口配置模式。

默认情况：所有端口默认属于 VLAN1。

使用指南：只有属于 Access mode 的端口才能加入到指定的 VLAN 中，并且 Access 端口同时只能加入到一个 VLAN 中。

举例：设置某 Access 端口加入 VLAN100。

DCRS-7604 (Config)#interface ethernet 1/8
DCRS-7604 (Config-ethernet1/8)#switchport mode access
DCRS-7604 (Config-ethernet1/8)#switchport access vlan 100
DCRS-7604 (Config-ethernet1/8)#exit

switchport interface

命令：switchport interface <interface-list>。

　　　no switchport interface <interface-list>。

功能：给 VLAN 分配以太网端口的命令；本命令的 no 操作为删除指定 VLAN 内的一个或一组端口。

参数：<interface-list>要添加或者删除的端口的列表，支持";"、"-"，如：ethernet 1/1;2;5 或 ethernet 1/1-6;8。

命令模式：VLAN 配置模式。

默认情况：新建立的 VLAN 默认不包含任何端口。

使用指南：Access 端口为普通端口，可以加入 VLAN，但同时只允许加入一个 VLAN。

举例：为 VLAN100 分配百兆以太网端口 1、3、4~7、8。

DCRS-7604 (Config-Vlan100)#switchport interface ethernet 1/1;3;4-7;8

switchport mode

命令：switchport mode {trunk|access}。

功能：设置交换机的端口为 access 模式或者 trunk 模式。

参数：trunk 表示端口允许通过多个 VLAN 流量；access 为端口只能属于一个 VLAN。

命令模式：接口配置模式。

默认情况：端口默认为 Access 模式。

使用指南：工作在 trunk mode 下的端口称为 Trunk 端口，Trunk 端口可以通过多个 VLAN 的流量，通过 Trunk 端口之间的互联，可以实现不同交换机上的相同 VLAN 的互通；工作在 access mode 下的端口称为 Access 端口，Access 端口可以分配给一个 VLAN，并且同时只能分配给一个 VLAN。

举例：将端口 5 设置为 trunk 模式，端口 8 设置为 access 模式。

DCRS-7604 (Config)#interface ethernet 1/5

DCRS-7604 (Config-ethernet1/5)#switchport mode trunk

DCRS-7604 (Config-ethernet1/5)#exit

DCRS-7604 (Config)#interface ethernet 1/8

DCRS-7604 (Config-ethernet1/8)#switchport mode access

DCRS-7604 (Config-ethernet1/8)#exit

switchport trunk allowed vlan

命令：switchport trunk allowed vlan {<vlan-list>|all}。

　　　　no switchport trunk allowed vlan。

功能：设置 Trunk 端口允许通过 VLAN；本命令的 no 操作为恢复默认情况。

参数：<vlan-list>为允许在该 Trunk 端口上通过的 VLAN 列表；all 关键字表示允许该 Trunk 端口通过所有 VLAN 的流量。

命令模式：接口配置模式。

默认情况：Trunk 端口默认允许通过所有 VLAN。

使用指南：用户可以通过本命令设置哪些 VLAN 的流量通过 Trunk 端口，没有包含的 VLAN 流量则被禁止。

举例：设置 Trunk 端口允许通过 VLAN1、3、5~20 的流量。

DCRS-7604 (Config)#interface ethernet 1/5
DCRS-7604 (Config-ethernet1/5)#switchport mode trunk
DCRS-7604 (Config-ethernet1/5)#switchport trunk allowed vlan 1;3;5-20
DCRS-7604 (Config-ethernet1/5)#exit

switchport trunk native vlan

命令：switchport trunk native vlan <vlan-id>。

no switchport trunk native vlan。

功能：设置 Trunk 端口的 PVID；本命令的 no 操作为恢复默认值。

参数：<vlan-id>为 Trunk 端口的 PVID。

命令模式：接口配置模式。

默认情况：Trunk 端口默认的 PVID 为 1。

使用指南：在 802.1Q 中定义了PVID这个概念。Trunk端口的PVID的作用是当一个untagged的帧进入Trunk端口，端口会对这个untagged帧打上带有本命令设置的native PVID的tag标记，用于VLAN的转发。

举例：设置某 Trunk 端口的 native vlan 为 100。

DCRS-7604 (Config)#interface ethernet 1/5
DCRS-7604 (Config-ethernet1/5)#switchport mode trunk
DCRS-7604 (Config-ethernet1/5)#switchport trunk native vlan 100
DCRS-7604 (Config-ethernet1/5)#exit

vlan ingress disable

命令：vlan ingress disable。

no vlan ingress disable。

功能：关闭端口的 VLAN 入口规则；本命令的 no 操作为打开入口规则。

命令模式：接口配置模式。

默认情况：系统默认打开端口的 VLAN 入口准则。

使用指南：当打开端口的 VLAN 入口规则，系统在接收数据时会检查源端口是否是该 VLAN 的成员端口，如果是则接收数据并转发到目的端口，否则丢弃该数据。

举例：关闭端口的 VLAN 入口规则。

DCRS-7604 (Config-Ethernet1/1)# vlan ingress disable

实训四　三层交换机实现二层交换机不同 VLAN 间通信

一、实训拓扑

该实训拓扑结构如图 1-44 所示。

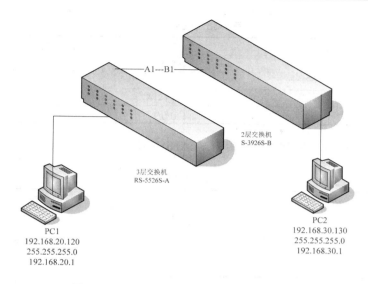

图 1-44　三层实现二层 Vlan 间通信实验拓扑

二、实训要求

如图 1-44 所示进行连线。

交换机 A 和 B 做如下配置：

1. 划分 vl 10(5-8)，vl 20(9-12)，vl 30(13-16)。

2. 设置 AB 交换机级联端口为 Trunk 模式。

3. 设置 3 层交换机 RS-5526S-A；启用三层路由功能；设置 vl10、vl20、vl30 的接口地址：int vl 10　192.168.10.1/24；int vl 20　192.168.20.1/24；int vl 30 192.168.30.1/24。

4. 根据连线位置正确配置 PC1、PC2 的地址(注意配上网关地址以及与 VLAN 的对应关系)。

实验结果：PC1---ping---PC2　　通。

三、实训步骤

步骤 1：对交换机 A 和 B 进行如下操作：

恢复出厂设置(Set default、write、reload)。

修改设备名称为：RS-A 和 S-B。

步骤 2：对 PC1 和 PC2 的本地连接 2 设置如下：

设备	IP 地址	gateway	Mask
PC1	192.168.20.120	192.168.20.1	255.255.255.0
PC2	192.168.30.130	192.168.30.1	255.255.255.0

注意：

将 PC1 和 PC2 的 "本地连接" 中的 "默认网关" 去掉。

步骤 3：在交换机 RS-A 上划分两个 VLAN，端口如表 1-38 所示。

表 1-38 VLAN 划分

VLAN	端口成员
10	5~8
20	9~12
30	13~16
Trunk 口	0/0/1

步骤 4：在交换机 S-B 上也划分两个 VLAN，端口如表 1-39 所示。

表 1-39 VLAN 划分

VLAN	端口成员
10	5~8
20	9~12
30	13~16
Trunk 口	0/0/1

步骤 5：交换机 A 的 1 口连交换机 B 的 1 口。

根据 PC 机网关与 VLAN 的对应关系，分析 PC 机的连线位置。

PC1 网关 192.168.20.1 对应 A-vlan 20 接口的 ip，所以 PC1 连接 A-vlan 20 的 9~12 位置。

PC2 网关 192.168.30.1 对应 A-vlan 30 接口的 ip，所以 PC1 连接 B-vlan 30 的 13~16 位置。

步骤 6：在交换机 A 上进行如下操作：

RS-A#Config
RS-A (Config)#interface vlan 10
RS-A (Config-If-Vlan10)#ip   address 192.168.10.1   255.255.255.0
RS-A (Config-If-Vlan10)#no shutdown
RS-A (Config-If-Vlan10)#exit
RS-A (Config)#interface vlan 20
RS-A (Config-If-Vlan20)#ip   address   192.168.20.1   255.255.255.0
RS-A (Config-If-Vlan20)#no shutdown
RS-A (Config-If-Vlan20)#exit
RS-A (Config)#interface vlan 30
RS-A (Config-If-Vlan30)#ip   address   192.168.30.1   255.255.255.0
RS-A (Config-If-Vlan30)#no shutdown
RS-A (Config-If-Vlan30)#exit

验证配置：

RS-A# show   ip   route
Total route items is 3, the matched route items is 3
Codes: C - connected, S - static, R - RIP derived, O - OSPF derived
       A - OSPF ASE,   B - BGP derived, D - DVMRP derived

Destination	Mask	Nexthop	Interface	Preference
C 192.168.10.0	255.255.255.0	0.0.0.0	Vlan10	0
C 192.168.20.0	255.255.255.0	0.0.0.0	Vlan20	0

C	192.168.30.0	255.255.255.0	0.0.0.0	Vlan30	0

步骤 7：对 PC1 和 PC2，进行如下验证实验：

"本地连接 2"都不配网关，PC1 和 PC2 彼此 ping 不通。

"本地连接 2"都配置网关，PC1 和 PC2 彼此能 ping 通。

实验十五　多层交换机静态路由实验

一、实验目的

1. 理解三层交换机进行路由的原理和具体实现拓扑。
2. 理解三层交换机静态路由的配置方法。

二、应用环境

当两台三层交换机级联时，为了保证每台交换机上所连接的网段可以和另一台交换机上连接的网段互相通信，最简单的方法就是设置静态路由。

三、实验设备

1. DCRS-7604(或 6804)交换机 1 台。
2. DCRS-5526S 交换机 1 台。
3. PC 机 2~4 台。
4. Console 线 1~2 根。
5. 直通网线 2~4 根。

四、实验拓扑

该实验拓扑结构如图 1-45 所示。

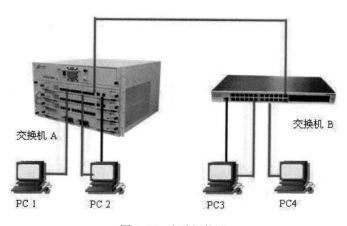

图 1-45　实验拓扑图

五、实验要求

1. 在交换机 A 和交换机 B 上分别划分基于端口的 VLAN。
2. 交换机 A 和 B 通过的 24 口级联。
3. 配置交换机 A 和 B 各 VLAN 虚拟接口的 IP 地址，如表 1-40、1-41 和 1-42 所示。

表 1-40 交换机 A/B 的 VLAN 划分

交换机	VLAN	端口成员
交换机 A	10	1~8
	20	9~16
	100	24
交换机 B	30	1~8
	40	9~16
	101	24

表 1-41 交换机 A/B 的 IP 配置

设备	IP 地址	gateway	Mask
PC1	192.168.10.101	192.168.10.1	255.255.255.0
PC2	192.168.20.101	192.168.20.1	255.255.255.0
PC3	192.168.30.101	192.168.30.1	255.255.255.0
PC4	192.168.40.101	192.168.40.1	255.255.255.0

表 1-42 PC 的配置

VLAN10	VLAN20	VLAN30	VLAN40	VLAN100	VLAN101
192.168.10.1	192.168.20.1	192.168.30.1	192.168.40.1	192.168.100.1	192.168.100.2

验证结果注意下列问题。

(1) 没有静态路由之前：

PC1 与 PC2，PC3 与 PC4 可以互通。

PC1、PC2 与 PC3、PC4 不通。

(2) 配置静态路由之后：

4 台 PC 之间都可以互通。

若实验结果和理论相符，则本实验完成。

六、实验步骤

第一步：交换机全部恢复出厂设置，配置交换机的 VLAN 信息。

交换机 A：

DCRS-7604#conf

DCRS-7604(Config)#vlan 10

DCRS-7604(Config-Vlan10)#switchport interface ethernet 1/1-8

Set the port Ethernet1/1 access vlan 10 successfully

Set the port Ethernet1/2 access vlan 10 successfully

Set the port Ethernet1/3 access vlan 10 successfully
Set the port Ethernet1/4 access vlan 10 successfully
Set the port Ethernet1/5 access vlan 10 successfully
Set the port Ethernet1/6 access vlan 10 successfully
Set the port Ethernet1/7 access vlan 10 successfully
Set the port Ethernet1/8 access vlan 10 successfully
DCRS-7604(Config-Vlan10)#exit
DCRS-7604(Config)#vlan 20
DCRS-7604(Config-Vlan20)#switchport interface ethernet 1/9-16
Set the port Ethernet1/9 access vlan 20 successfully
Set the port Ethernet1/10 access vlan 20 successfully
Set the port Ethernet1/11 access vlan 20 successfully
Set the port Ethernet1/12 access vlan 20 successfully
Set the port Ethernet1/13 access vlan 20 successfully
Set the port Ethernet1/14 access vlan 20 successfully
Set the port Ethernet1/15 access vlan 20 successfully
Set the port Ethernet1/16 access vlan 20 successfully
DCRS-7604(Config-Vlan20)#exit
DCRS-7604(Config)#vlan 100
DCRS-7604(Config-Vlan100)#switchport interface ethernet 1/24
Set the port Ethernet1/24 access vlan 100 successfully
DCRS-7604(Config-Vlan100)#exit
DCRS-7604(Config)#

验证配置：

DCRS-7604#show vlan

VLAN	Name	Type	Media	Ports	
1	default	Static	ENET	Ethernet1/17	Ethernet1/18
				Ethernet1/19	Ethernet1/20
				Ethernet1/21	Ethernet1/22
				Ethernet1/23	Ethernet1/24
				Ethernet1/25	Ethernet1/26
				Ethernet1/27	
10	VLAN0010	Static	ENET	Ethernet1/1	Ethernet1/2
				Ethernet1/3	Ethernet1/4
				Ethernet1/5	Ethernet1/6
				Ethernet1/7	Ethernet1/8
20	VLAN0020	Static	ENET	Ethernet1/9	Ethernet1/10
				Ethernet1/11	Ethernet1/12
				Ethernet1/13	Ethernet1/14
				Ethernet1/15	Ethernet1/16
100	VLAN0100	Static	ENET	Ethernet1/24	

DCRS-7604#

交换机 B：

DCRS-5526S(Config)#vlan 30
DCRS-5526S(Config-Vlan30)#switchport interface ethernet 0/0/1-8
Set the port Ethernet0/0/1 access vlan 30 successfully
Set the port Ethernet0/0/2 access vlan 30 successfully
Set the port Ethernet0/0/3 access vlan 30 successfully
Set the port Ethernet0/0/4 access vlan 30 successfully
Set the port Ethernet0/0/5 access vlan 30 successfully
Set the port Ethernet0/0/6 access vlan 30 successfully
Set the port Ethernet0/0/7 access vlan 30 successfully
Set the port Ethernet0/0/8 access vlan 30 successfully
DCRS-5526S(Config-Vlan30)#exit
DCRS-5526S(Config)#vlan 40
DCRS-5526S(Config-Vlan40)#switchport interface ethernet 0/0/9-16
Set the port Ethernet0/0/9 access vlan 40 successfully
Set the port Ethernet0/0/10 access vlan 40 successfully
Set the port Ethernet0/0/11 access vlan 40 successfully
Set the port Ethernet0/0/12 access vlan 40 successfully
Set the port Ethernet0/0/13 access vlan 40 successfully
Set the port Ethernet0/0/14 access vlan 40 successfully
Set the port Ethernet0/0/15 access vlan 40 successfully
Set the port Ethernet0/0/16 access vlan 40 successfully
DCRS-5526S(Config-Vlan40)#exit
DCRS-5526S(Config)#vlan 101
DCRS-5526S(Config-Vlan101)#switchport interface ethernet 0/0/24
Set the port Ethernet0/0/24 access vlan 101 successfully
DCRS-5526S(Config-Vlan101)#exit
DCRS-5526S(Config)#

验证配置：

DCRS-5526S#show vlan

VLAN	Name	Type	Media	Ports	
1	default	Static	ENET	Ethernet0/0/17	Ethernet0/0/18
				Ethernet0/0/19	Ethernet0/0/20
				Ethernet0/0/21	Ethernet0/0/22
				Ethernet0/0/23	
30	VLAN0030	Static	ENET	Ethernet0/0/1	Ethernet0/0/2
				Ethernet0/0/3	Ethernet0/0/4
				Ethernet0/0/5	Ethernet0/0/6
				Ethernet0/0/7	Ethernet0/0/8
40	VLAN0040	Static	ENET	Ethernet0/0/9	Ethernet0/0/10
				Ethernet0/0/11	Ethernet0/0/12

				Ethernet0/0/13	Ethernet0/0/14
				Ethernet0/0/15	Ethernet0/0/16
101	VLAN0101	Static	ENET	Ethernet0/0/24	

DCRS-5526S#

第二步：配置交换机各 VLAN 虚接口的 IP 地址。

交换机 A：

DCRS-7604(Config)#int vlan 10
DCRS-7604(Config-If-Vlan10)#ip address 192.168.10.1 255.255.255.0
DCRS-7604(Config-If-Vlan10)#no shut
DCRS-7604(Config-If-Vlan10)#exit
DCRS-7604(Config)#int vlan 20
DCRS-7604(Config-If-Vlan20)#ip address 192.168.20.1 255.255.255.0
DCRS-7604(Config-If-Vlan20)#no shut
DCRS-7604(Config-If-Vlan20)#exit
DCRS-7604(Config)#int vlan 100
DCRS-7604(Config-If-Vlan100)#ip address 192.168.100.1 255.255.255.0
DCRS-7604(Config-If-Vlan100)#no shut
DCRS-7604(Config-If-Vlan100)#
DCRS-7604(Config-If-Vlan100)#exit
DCRS-7604(Config)#

交换机 B：

DCRS-5526S(Config)#int vlan 30
DCRS-5526S(Config-If-Vlan30)#ip address 192.168.30.1 255.255.255.0
DCRS-5526S(Config-If-Vlan30)#no shut
DCRS-5526S(Config-If-Vlan30)#exit
DCRS-5526S(Config)#interface vlan 40
DCRS-5526S(Config-If-Vlan40)#ip address 192.168.40.1 255.255.255.0
DCRS-5526S(Config-If-Vlan40)#exit
DCRS-5526S(Config)#int vlan 101
DCRS-5526S(Config-If-Vlan101)#ip address 192.168.100.2 255.255.255.0
DCRS-5526S(Config-If-Vlan101)#exit
DCRS-5526S(Config)#

第三步：如表 1-43 所示，配置各 PC 的 IP 地址，注意配置网关。

表 1-43　PC 的配置

设备	IP 地址	gateway	Mask
PC1	192.168.10.101	192.168.10.1	255.255.255.0
PC2	192.168.20.101	192.168.20.1	255.255.255.0
PC3	192.168.30.101	192.168.30.1	255.255.255.0
PC4	192.168.40.101	192.168.40.1	255.255.255.0

第四步：如表 1-44 所示，验证 PC 之间的连通性。

表 1-44　验证 PC 之间的连通性

PC	端口	PC	端口	结果	原因
PC1	A：1/1	PC2	A：1/9	通	
PC1	A：1/1	Vlan 100	A：1/24	通	
PC1	A：1/1	Vlan 101	B：0/0/24	不通	
PC1	A：1/1	PC3	B：0/0/1	不通	

查看路由表，进一步分析上一步的现象原因。

交换机 A：

DCRS-7604#show ip route

Total route items is 3, the matched route items is 3

Codes: C - connected, S - static, R - RIP derived, O - OSPF derived

　　　　A - OSPF ASE, B - BGP derived, D - DVMRP derived

Destination	Mask	Nexthop	Interface	Preference
C　192.168.10.0	255.255.255.0	0.0.0.0	Vlan10	0
C　192.168.20.0	255.255.255.0	0.0.0.0	Vlan20	0
C　192.168.100.0	255.255.255.0	0.0.0.0	Vlan100	0

DCRS-7604#

交换机 B：

DCRS-5526S#show ip route

Total route items is 3, the matched route items is 3

Codes: C - connected, S - static, R - RIP derived, O - OSPF derived

　　　　A - OSPF ASE, B - BGP derived, D - DVMRP derived

Destination	Mask	Nexthop	Interface	Preference
C　192.168.30.0	255.255.255.0	0.0.0.0	Vlan30	0
C　192.168.40.0	255.255.255.0	0.0.0.0	Vlan40	0
C　192.168.100.0	255.255.255.0	0.0.0.0	Vlan101	0

DCRS-5526S#

第五步：配置静态路由。

交换机 A：

DCRS-7604(Config)#ip route 192.168.30.0 255.255.255.0 192.168.100.2

DCRS-7604(Config)#ip route 192.168.40.0 255.255.255.0 192.168.100.2

验证配置：

DCRS-7604#show ip route

Total route items is 5, the matched route items is 5

Codes: C - connected, S - static, R - RIP derived, O - OSPF derived

　　　　A - OSPF ASE, B - BGP derived, D - DVMRP derived

Destination	Mask	Nexthop	Interface	Preference
C 192.168.10.0	255.255.255.0	0.0.0.0	Vlan10	0
C 192.168.20.0	255.255.255.0	0.0.0.0	Vlan20	0
S 192.168.30.0	255.255.255.0	192.168.100.2	Vlan100	1
S 192.168.40.0	255.255.255.0	192.168.100.2	Vlan100	1
C 192.168.100.0	255.255.255.0	0.0.0.0	Vlan100	0

DCRS-7604#

交换机 B：

DCRS-5526S(Config)#ip route 192.168.10.0 255.255.255.0 192.168.100.1
DCRS-5526S(Config)#ip route 192.168.20.0 255.255.255.0 192.168.100.1

验证配置：

DCRS-5526S#show ip route
Total route items is 5, the matched route items is 5
Codes: C - connected, S - static, R - RIP derived, O - OSPF derived
       A - OSPF ASE, B - BGP derived, D - DVMRP derived

Destination	Mask	Nexthop	Interface	Preference
S 192.168.10.0	255.255.255.0	192.168.100.1	Vlan101	1
S 192.168.20.0	255.255.255.0	192.168.100.1	Vlan101	1
C 192.168.30.0	255.255.255.0	0.0.0.0	Vlan30	0
C 192.168.40.0	255.255.255.0	0.0.0.0	Vlan40	0
C 192.168.100.0	255.255.255.0	0.0.0.0	Vlan101	0

DCRS-5526S#

第六步：如表 1-45 所示，验证 PC 之间的连通。

表 1-45 验证 PC 之间的连通

PC	端口	PC	端口	结果	原因
PC1	A：1/1	PC2	A：1/9	通	
PC1	A：1/1	Vlan 100	A：1/24	通	
PC1	A：1/1	Vlan 101	B：0/0/24	通	
PC1	A：1/1	PC3	B：0/0/1	通	

七、注意事项和排错

1. PC 机一定要配置正确的网关，否则不能正常通信。
2. 两台交换机级联的端口可以在同一 VLAN，也可以在不同的 VLAN。

八、思考题

1. 如果把交换机 B 上的 vlan30 改成 vlan10，请问两台交换机上的 vlan10 是同一个 VLAN 么？

2. 第四步中，PC1 ping vlan101 以及 PC1 ping PC3 都不通，其原因各是什么？

实训五　多层交换机的静态路由配置

一、实训设备

(1) DCS-3926S 交换机　　2 台。
(2) DCRS-5526S 交换机　　2 台。
(3) PC 机　　4 台。
(4) 直通网线　　7 根。

二、实训拓扑

该实验拓扑结构如图 1-46 所示。

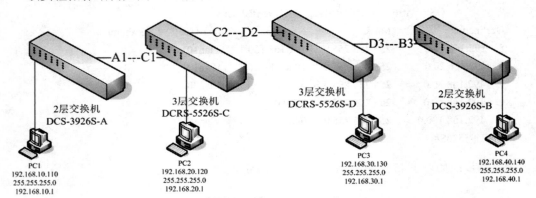

图 1-46　多层交换机的静态路由实验拓扑

三、实训要求

如图 1-46 所示进行连线。
实训结果：PC1---ping---PC2----PC3----PC4　　全通。

四、实训步骤

1. 交换机 A、B、C、D 做如下配置：
A 划分 vl 10(9-12)，vl 20(13-16)，Trunk (1)；
B 划分 vl 30(17-20)，vl 40(21-24)，Trunk (3)；
C 划分 vl 10(9-12)，vl 20(13-16)，vl 100(2)，Trunk (1)；
D 划分 vl 30(17-20)，vl 40(21-24)，vl 101(2)，Trunk (3)。
2. 设置 3 层交换机 C 启用三层路由功能。

设置 vl 10、vl 20、vl 100 的接口地址：

int vl 10　　192.168.10.1/24；
int vl 20　　192.168.20.1/24；
int vl 100 192.168.100.1/30

设置 3 层交换机 D 启用三层路由功能。
设置 vl 30、vl 40、vl 101 的接口地址：

int vl 30　　192.168.30.1/24　；
int vl 40　　192.168.40.1/24；
int vl 101 192.168.100.2/30

3. 配置 C 的静态路由。

C(config)#ip route 192.168.30.0 255.255.255.0 192.168.100.2
C(config)#ip route 192.168.40.0 255.255.255.0 192.168.100.2

配置 D 的静态路由。

D(config)#ip route 192.168.10.0 255.255.255.0 192.168.100.1
D(config)#ip route 192.168.20.0 255.255.255.0 192.168.100.1

4. 根据连线位置正确配置 PC1、PC2、PC3、PC4 的地址(注意配上网关地址以及与 vlan 的对应关系)。

实验结果：PC1---ping---PC2----PC3----PC4　　　全通。
查看交换机状态：sh vlan　　　　sh run　　　　sh ip route。

实验十六　三层交换机 RIP 动态路由

一、实验目的

1. 掌握三层交换机之间通过 RIP 协议实现网段互通的配置方法。
2. 理解动态实现方式与静态方式的不同。

二、应用环境

当两台三层交换机级联时，为了保证每台交换机上所连接的网段可以和另一台交换机上连接的网段互相通信，使用 RIP 协议可以动态学习路由。

三、实验设备

1. DCRS-7604(或 6804)交换机　　　1 台。

2. DCRS-5526S 交换机   1 台。
3. PC 机   2~4 台。
4. Console 线   1~2 根。
5. 直通网线   2~4 根。

四、实验拓扑

该实验拓扑结构如图 1-47 所示。

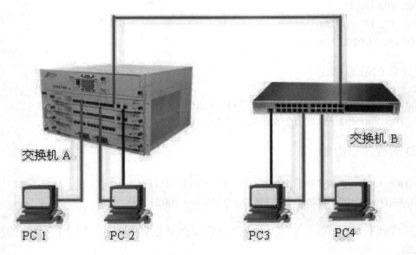

图 1-47 实验拓扑图

五、实验要求

1. 如表 1-46 所示，在交换机 A 和交换机 B 上分别划分基于端口的 VLAN。

表 1-46 交换机 A/B 的 VLAN 划分

交换机	VLAN	端口成员
交换机 A	10	1~8
	20	9~16
	100	24
交换机 B	30	1~8
	40	9~16
	101	24

2. 交换机 A 和 B 通过的 24 口级联。
3. 按如表 1-47 所示配置交换机 A 和 B 各 VLAN 虚拟接口的 IP 地址。

表 1-47 交换机 A/B 的 IP 配置

VLAN10	VLAN20	VLAN30	VLAN40	VLAN100	VLAN101
192.168.10.1	192.168.20.1	192.168.30.1	192.168.40.1	192.168.100.1	192.168.100.2

4. PC 的配置如表 1-48 所示。

表 1-48 PC 的配置

设备	IP 地址	gateway	Mask
PC1	192.168.10.101	192.168.10.1	255.255.255.0
PC2	192.168.20.101	192.168.20.1	255.255.255.0
PC3	192.168.30.101	192.168.30.1	255.255.255.0
PC4	192.168.40.101	192.168.40.1	255.255.255.0

5. 验证。

没有 RIP 路由协议之前：

PC1 与 PC2、PC3 与 PC4 可以互通。

PC1、PC2 与 PC3、PC4 不通。

配置 RIP 路由协议之后：

4 台 PC 之间都可以互通。

若实验结果和理论相符，则本实验完成。

六、实验步骤

第一步：交换机全部恢复出厂设置，配置交换机的 VLAN 信息。

交换机 A：

DCRS-7604#conf
DCRS-7604(Config)#vlan 10
DCRS-7604(Config-Vlan10)#switchport interface ethernet 1/1-8
Set the port Ethernet1/1 access vlan 10 successfully
Set the port Ethernet1/2 access vlan 10 successfully
Set the port Ethernet1/3 access vlan 10 successfully
Set the port Ethernet1/4 access vlan 10 successfully
Set the port Ethernet1/5 access vlan 10 successfully
Set the port Ethernet1/6 access vlan 10 successfully
Set the port Ethernet1/7 access vlan 10 successfully
Set the port Ethernet1/8 access vlan 10 successfully
DCRS-7604(Config-Vlan10)#exit
DCRS-7604(Config)#vlan 20
DCRS-7604(Config-Vlan20)#switchport interface ethernet 1/9-16
Set the port Ethernet1/9 access vlan 20 successfully
Set the port Ethernet1/10 access vlan 20 successfully
Set the port Ethernet1/11 access vlan 20 successfully
Set the port Ethernet1/12 access vlan 20 successfully
Set the port Ethernet1/13 access vlan 20 successfully
Set the port Ethernet1/14 access vlan 20 successfully
Set the port Ethernet1/15 access vlan 20 successfully

Set the port Ethernet1/16 access vlan 20 successfully
DCRS-7604(Config-Vlan20)#exit
DCRS-7604(Config)#vlan 100
DCRS-7604(Config-Vlan100)#switchport interface ethernet 1/24
Set the port Ethernet1/24 access vlan 100 successfully
DCRS-7604(Config-Vlan100)#exit
DCRS-7604(Config)#

验证配置：

DCRS-7604#show vlan

VLAN	Name	Type	Media	Ports	
1	default	Static	ENET	Ethernet1/17	Ethernet1/18
				Ethernet1/19	Ethernet1/20
				Ethernet1/21	Ethernet1/22
				Ethernet1/23	Ethernet1/24
				Ethernet1/25	Ethernet1/26
				Ethernet1/27	
10	VLAN0010	Static	ENET	Ethernet1/1	Ethernet1/2
				Ethernet1/3	Ethernet1/4
				Ethernet1/5	Ethernet1/6
				Ethernet1/7	Ethernet1/8
20	VLAN0020	Static	ENET	Ethernet1/9	Ethernet1/10
				Ethernet1/11	Ethernet1/12
				Ethernet1/13	Ethernet1/14
				Ethernet1/15	Ethernet1/16
100	VLAN0100	Static	ENET	Ethernet1/24	

DCRS-7604#

交换机 B：

DCRS-5526S(Config)#vlan 30
DCRS-5526S(Config-Vlan30)#switchport interface ethernet 0/0/1-8
Set the port Ethernet0/0/1 access vlan 30 successfully
Set the port Ethernet0/0/2 access vlan 30 successfully
Set the port Ethernet0/0/3 access vlan 30 successfully
Set the port Ethernet0/0/4 access vlan 30 successfully
Set the port Ethernet0/0/5 access vlan 30 successfully
Set the port Ethernet0/0/6 access vlan 30 successfully
Set the port Ethernet0/0/7 access vlan 30 successfully
Set the port Ethernet0/0/8 access vlan 30 successfully
DCRS-5526S(Config-Vlan30)#exit
DCRS-5526S(Config)#vlan 40
DCRS-5526S(Config-Vlan40)#switchport interface ethernet 0/0/9-16
Set the port Ethernet0/0/9 access vlan 40 successfully

Set the port Ethernet0/0/10 access vlan 40 successfully
Set the port Ethernet0/0/11 access vlan 40 successfully
Set the port Ethernet0/0/12 access vlan 40 successfully
Set the port Ethernet0/0/13 access vlan 40 successfully
Set the port Ethernet0/0/14 access vlan 40 successfully
Set the port Ethernet0/0/15 access vlan 40 successfully
Set the port Ethernet0/0/16 access vlan 40 successfully
DCRS-5526S(Config-Vlan40)#exit
DCRS-5526S(Config)#vlan 101
DCRS-5526S(Config-Vlan101)#switchport interface ethernet 0/0/24
Set the port Ethernet0/0/24 access vlan 101 successfully
DCRS-5526S(Config-Vlan101)#exit
DCRS-5526S(Config)#

验证配置：

DCRS-5526S#show vlan

VLAN	Name	Type	Media	Ports	
1	default	Static	ENET	Ethernet0/0/17	Ethernet0/0/18
				Ethernet0/0/19	Ethernet0/0/20
				Ethernet0/0/21	Ethernet0/0/22
				Ethernet0/0/23	
30	VLAN0030	Static	ENET	Ethernet0/0/1	Ethernet0/0/2
				Ethernet0/0/3	Ethernet0/0/4
				Ethernet0/0/5	Ethernet0/0/6
				Ethernet0/0/7	Ethernet0/0/8
40	VLAN0040	Static	ENET	Ethernet0/0/9	Ethernet0/0/10
				Ethernet0/0/11	Ethernet0/0/12
				Ethernet0/0/13	Ethernet0/0/14
				Ethernet0/0/15	Ethernet0/0/16
101	VLAN0101	Static	ENET	Ethernet0/0/24	

DCRS-5526S#

第二步：配置交换机各 VLAN 虚接口的 IP 地址。

交换机 A：

DCRS-7604(Config)#int vlan 10
DCRS-7604(Config-If-Vlan10)#ip address 192.168.10.1 255.255.255.0
DCRS-7604(Config-If-Vlan10)#no shut
DCRS-7604(Config-If-Vlan10)#exit
DCRS-7604(Config)#int vlan 20
DCRS-7604(Config-If-Vlan20)#ip address 192.168.20.1 255.255.255.0
DCRS-7604(Config-If-Vlan20)#no shut
DCRS-7604(Config-If-Vlan20)#exit

```
DCRS-7604(Config)#int vlan 100
DCRS-7604(Config-If-Vlan100)#ip address 192.168.100.1 255.255.255.0
DCRS-7604(Config-If-Vlan100)#no shut
DCRS-7604(Config-If-Vlan100)#
DCRS-7604(Config-If-Vlan100)#exit
DCRS-7604(Config)#
```

交换机 B：

```
DCRS-5526S(Config)#int vlan 30
DCRS-5526S(Config-If-Vlan30)#ip address 192.168.30.1 255.255.255.0
DCRS-5526S(Config-If-Vlan30)#no shut
DCRS-5526S(Config-If-Vlan30)#exit
DCRS-5526S(Config)#interface vlan 40
DCRS-5526S(Config-If-Vlan40)#ip address 192.168.40.1 255.255.255.0
DCRS-5526S(Config-If-Vlan40)#exit
DCRS-5526S(Config)#int vlan 101
DCRS-5526S(Config-If-Vlan101)#ip address 192.168.100.2 255.255.255.0
DCRS-5526S(Config-If-Vlan101)#exit
DCRS-5526S(Config)#
```

第三步：按如表 1-49 所示配置各 PC 的 IP 地址，注意配置网关。

表 1-49 PC 的配置

设备	IP 地址	gateway	Mask
PC1	192.168.10.101	192.168.10.1	255.255.255.0
PC2	192.168.20.101	192.168.20.1	255.255.255.0
PC3	192.168.30.101	192.168.30.1	255.255.255.0
PC4	192.168.40.101	192.168.40.1	255.255.255.0

第四步：如表 1-50 所示，验证 PC 之间的连通性。

表 1-50 PC 的配置

PC	端口	PC	端口	结果	原因
PC1	A：1/1	PC2	A：1/9	通	
PC1	A：1/1	Vlan 100	A：1/24	通	
PC1	A：1/1	Vlan 101	B：0/0/24	不通	
PC1	A：1/1	PC3	B：0/0/1	不通	

查看路由表，进一步分析上一步的现象原因。

交换机 A：

```
DCRS-7604#show ip route
Total route items is 3, the matched route items is 3
Codes: C - connected, S - static, R - RIP derived, O - OSPF derived
```

A - OSPF ASE, B - BGP derived, D - DVMRP derived

Destination	Mask	Nexthop	Interface	Preference
C 192.168.10.0	255.255.255.0	0.0.0.0	Vlan10	0
C 192.168.20.0	255.255.255.0	0.0.0.0	Vlan20	0
C 192.168.100.0	255.255.255.0	0.0.0.0	Vlan100	0

DCRS-7604#

交换机 B：

DCRS-5526S#show ip route
Total route items is 3, the matched route items is 3
Codes: C - connected, S - static, R - RIP derived, O - OSPF derived
A - OSPF ASE, B - BGP derived, D - DVMRP derived

Destination	Mask	Nexthop	Interface	Preference
C 192.168.30.0	255.255.255.0	0.0.0.0	Vlan30	0
C 192.168.40.0	255.255.255.0	0.0.0.0	Vlan40	0
C 192.168.100.0	255.255.255.0	0.0.0.0	Vlan101	0

DCRS-5526S#

第五步：启动 RIP 协议，并将对应的直连网段配置到 RIP 进程中。

交换机 A：

DCRS-7604(Config)#router rip
DCRS-7604(Config-Router-Rip)#version 2
DCRS-7604(Config)#interface vlan 10
DCRS-7604(Config-If-Vlan10)#ip rip work
DCRS-7604(Config-If-Vlan10)#exit
DCRS-7604(Config)#interface vlan 20
DCRS-7604(Config-If-Vlan20)#ip rip work
DCRS-7604(Config-If-Vlan20)#exit
DCRS-7604(Config)#interface vlan 100
DCRS-7604(Config-If-Vlan100)#ip rip work
DCRS-7604(Config-If-Vlan100)#exit

验证配置：

DCRS-7604#show ip rip
RIP information:
Automatic network summarization is not in effect.
default metric for redistribute is :1.
neighbour is :NULL
preference is :120
RIP version information is :

interface	send version	receive version
Vlan10	V2MC	V2
Vlan20	V2MC	V2

Vlan100   V2MC   V2
DCRS-7604#

DCRS-7604#show ip route

Total route items is 4, the matched route items is 4

Codes: C - connected, S - static, R - RIP derived, O - OSPF derived
       A - OSPF ASE, B - BGP derived, D - DVMRP derived

	Destination	Mask	Nexthop	Interface	Preference
C	192.168.10.0	255.255.255.0	0.0.0.0	Vlan10	0
R	192.168.30.0	255.255.255.0	192.168.100.2	Vlan100	120
R	192.168.40.0	255.255.255.0	192.168.100.2	Vlan100	120
C	192.168.100.0	255.255.255.0	0.0.0.0	Vlan100	0

DCRS-7604#

(R 表示 rip 协议学习到的网段)

交换机 B：

DCRS-5526S(Config)#router rip
DCRS-5526S(Config-Router-Rip)#version 2
DCRS-5526S(Config-Router-Rip)#exit
DCRS-5526S(Config)#interface vlan 30
DCRS-5526S(Config-If-Vlan30)#ip rip work
DCRS-5526S(Config-If-Vlan30)#exit
DCRS-5526S(Config)#interface vlan 40
DCRS-5526S(Config-If-Vlan40)#ip rip work
DCRS-5526S(Config-If-Vlan40)#exit
DCRS-5526S(Config)#interface vlan 101
DCRS-5526S(Config-If-Vlan101)#ip rip work
DCRS-5526S(Config-If-Vlan101)#exit
DCRS-5526S(Config)#

验证配置：

DCRS-5526S#show ip rip
RIP information:
Automatic network summarization is not in effect.
default metric for redistribute is :1.
neighbour is :NULL
preference is :120
RIP version information is :

interface	send version	receive version
Vlan30	V2MC	V2
Vlan40	V2MC	V2
Vlan101	V2MC	V2

DCRS-5526S#
DCRS-5526S#show ip route

Total route items is 4, the matched route items is 4
Codes: C - connected, S - static, R - RIP derived, O - OSPF derived
A - OSPF ASE, B - BGP derived, D - DVMRP derived

	Destination	Mask	Nexthop	Interface	Preference
R	192.168.10.0	255.255.255.0	192.168.100.1	Vlan101	120
C	192.168.30.0	255.255.255.0	0.0.0.0	Vlan30	0
C	192.168.40.0	255.255.255.0	0.0.0.0	Vlan40	0
C	192.168.100.0	255.255.255.0	0.0.0.0	Vlan101	0

DCRS-5526S#

第六步：如表 1-51 所示，验证 PC 之间的连通。

表 1-51 验证 PC 之间的连通

PC	端口	PC	端口	结果	原因
PC1	A：1/1	PC2	A：1/9	通	
PC1	A：1/1	Vlan 100	A：1/24	通	
PC1	A：1/1	Vlan 101	B：0/0/24	通	
PC1	A：1/1	PC3	B：0/0/1	通	

七、注意事项和排错

1. 全局启动"router rip"之后，交换机自动会在所有的虚接口上启动 rip 协议。
2. 可以在单个虚接口上禁止 rip 协议。

八、思考题

1. 如果在交换机 A 的 vlan100 上禁止 rip 协议，请问 PC1 还能 ping 通 PC3 么？
2. 如果在交换机 B 的 vlan30 上禁止 rip 协议，请问 PC1 还能 ping 通 PC3 么？

九、相关配置命令详解

RIP 的配置命令：

auto-summary
default-metric
ip rip authentication key-chain
ip rip authentication mode
ip rip metricin
ip rip metricout
ip rip input
ip rip output
ip rip receive version
ip rip send version
ip rip work
ip split horizon

redistribute
rip broadcast
rip checkzero
rip preference
router rip
timer basic
version
show ip protocols
show ip rip
debug ip rip packet
debug ip rip recv
debug ip rip send

auto-summary

命令：auto-summary。

no auto-summary。

功能：配置路由聚合功能，本命令的 no 操作取消路由聚合功能。

参数：无。

默认情况：不使用自动路由聚合功能。

命令模式：RIP 协议配置模式。

使用指南：路由聚合减少了在路由表中的路由信息量，也减少了交换信息量。RIP-1 不支持子网掩码，如果转发子网路由有可能会引起歧义。所以，RIP-1 始终启用路由聚合功能。如果使用 RIP-2，可以通过 no auto-summary 命令关闭路由聚合功能。当需要将子网路由广播出去时，可以关闭路由聚合功能。

举例：将 RIP 版本设为 RIP-2 并关闭路由聚合功能。

Switch(Config)#router rip
Switch(Config-Router-Rip(下同))#version 2
Switch(config-router-rip)#no auto-summary

相关命令：version。

default-metric

命令：default-metric <value>。

no default-metric。

功能：设定引入路由的默认路由权值；本命令的 no 操作为恢复默认值。

参数：<value>为所要设定的路由权值，取值范围 1~16。

默认情况：默认的路由权值为 1。

命令模式：RIP 协议配置模式。

使用指南：default-metric 命令用于设定将其他路由协议的路由引入到 RIP 路由时使用的默认路由权值。当使用 redistribute 命令引入其他协议的路由时，如果不指定具体的路由权值，则以 default-metric 所指定的默认路由权值引入。

举例：设定在 RIP 路由中引入其他路由协议的默认路由权值为 3。

Switch(config-router-rip)#default-metric 3

相关命令：redistribute。

ip rip authentication key-chain

命令：ip rip authentication key-chain <name-of-chain>。
　　　no ip rip authentication key-chain。

功能：设置 RIP 验证使用的密钥；本命令 no 操作为取消对 RIP 验证。

参数：<name-of-chain>为字符串，最长不超过 16 个字符。

默认情况：系统默认不进行 RIP 验证。

命令模式：接口配置模式。

使用指南：本命令的 no 操作为取消 RIP 验证，而不是删除 RIP 验证时使用的密钥。

相关命令：ip rip authentication。

ip rip authentication mode

命令：ip rip authentiaction mode {text|md5 type {cisco|usual}}。
　　　no ip rip authentication mode。

功能：设置验证使用的类型；本命令的 no 操作为恢复默认验证类型即文本验证。

参数：text 表示文本验证；md5 表示 MD5 验证，并且 MD5 验证又分为 Cisco MD5 和常规 MD5 两种验证方法。

默认情况：默认使用文本验证。

命令模式：接口配置模式。

使用指南：RIP-I 不支持验证，RIP-II 支持两种验证：文本验证(即 Simple 验证)和数据报验证(即 MD5 验证)。MD5 验证的数据报格式有两种：一种遵循 RFC1723(RIP Version 2Carrying Additional Information)规定，另一种遵循 RFC2082(RIP-II MD5 Authentication)规定。

举例：在接口 vlan1 上设置 RIP 报文的 Cisco MD5 验证，验证使用的密钥为 digitalchina。

Switch(config-If-Vlan1)#ip rip authentication mode md5 type cisco
Switch(config-If-Vlan1)#ip rip authentication key-chain digitalchina

相关命令：ip rip authentication key-chain。

ip rip metricin

命令：ip rip metricin <value>。
　　　no ip rip metricin。

功能：设置在接口接收 RIP 报文增加的附加路由权值；本命令的 no 操作为恢复默认值。

参数：<value>为附加的路由权值，取值范围 1~15。

默认情况：RIP 在接收报文时默认的附加路由权值为 1。

命令模式：接口配置模式。

相关命令：ip rip metricout。

ip rip metricout

命令：ip rip metricout <value>。

　　　　no ip rip metricout。

功能：设置从接口发送 RIP 报文增加的附加路由权值；本命令的 no 操作为恢复默认值。

参数：<value>为附加的路由权值，取值范围是 0~15。

默认情况：RIP 在发送报文时默认的附加路由权值为 0。

命令模式：接口配置模式。

举例：在接口 vlan1 接收 RIP 报文附加路由权值 5，发送 RIP 报文附加路由权值 3。

```
Switch(config-If-Vlan1)#ip rip metricin 5
Switch(config-If-Vlan1)#ip rip metricout 3
```

相关命令：ip rip metricin。

ip rip input

命令：ip rip input。

　　　　no ip rip input。

功能：设置接口能够接收 RIP 报文；本命令的 no 操作为接口不能接收 RIP 报文。

默认情况：接口默认为接收 RIP 报文。

命令模式：接口配置模式。

使用指南：本命令是与其他两条命令 ip rip output 和 ip rip work 协作使用的，ip rip work 从功能上等价于 ip rip input & ip rip output，后两条命令分别控制接口上对 RIP 报文的接收和发送，前一项命令等于后两条命令作用之和。

相关命令：ip rip output。

ip rip output

命令：ip rip output。

　　　　no ip rip output。

功能：设置接口能够向外发送 RIP 报文；本命令的 no 操作为接口不能向外发送 RIP 报文。

默认情况：接口默认向外发送 RIP 报文。

命令模式：接口配置模式。

使用指南：本命令是与其他两条命令 ip rip output 和 ip rip work 协作使用的，ip rip work 从功能上等价于 ip rip input & ip rip output，后两条命令分别控制接口上对 RIP 报文的接收和发送，前一项命令等于后两条命令作用之和。

相关命令：ip rip input。

ip rip receive version

命令：ip rip receive version {v1 | v2 | v12}。

　　　　no ip rip receive version。

功能：设置接口接收 RIP 报文的版本信息。默认接收 RIP 版本 1 和 2；本命令的 no 操作恢复默认值。

参数：v1 和 v2 表示 RIP 版本 1 和 RIP 版本 2，v12 表示 RIP 版本 1、2。

默认情况：默认为 v12，即 RIP 版本 1 和 2 全收。

命令模式：接口配置模式。

ip rip send version

命令：ip rip send version {v1 | v2 [bc|mc]}。

　　　　no ip rip send version。

功能：设置接口发送 RIP 报文的版本；本命令的 no 操作为恢复默认值。

参数：v1 | v2 均为 RIP 的版本号；[bc|mc] 只在发送 RIP 版本 2 的情况下配置，用于指定发送方式，BC 为广播方式，MC 为组播方式。当配置发送 RIP 版本 2 报文时，接口默认以组播 MC 方式发送 RIP 版本 2 报文，只有在设置 BC 后才能在此接口发送广播报文。

默认情况：接口默认发送 RIP 版本 2 报文。

命令模式：接口配置模式。

使用指南：当配置接口发送 RIP 版本 2 报文时，默认发送方式为组播方式，只有设置了 BC 方式后才能在此接口发送广播报文。

ip rip work

命令：ip rip work。

　　　　no ip rip work。

功能：设置接口上是否运行 RIP 协议；本命令的 no 操作为本接口不收发 RIP 报文。

默认情况：打开 RIP 路由开关后，接口上默认为运行 RIP 协议。

命令模式：接口配置模式。

使用指南：本命令在功能上等价于 ip rip input & ip rip output，后两条命令分别控制接口上对 RIP 报文的接收和发送，前一项命令等于后两条命令作用之和。

相关命令：ip rip input、ip rip output。

ip split-horizon

命令：ip split-horizon。

　　　　no ip split-horizon。

功能：设置允许水平分割；本命令的 no 操作为禁止水平分割。

默认情况：默认情况下，允许水平分割。

命令模式：接口配置模式。

使用指南：水平分割用于防止路由循环(Routing Loops)，即防止三层交换机把从一个接口上所学习到的路由再从这个接口广播出去。

举例：在接口 vlan1 上设置禁止水平分割。

Switch(config)#interface vlan1
Switch(config-If-Vlan1)#no ip split-horizon

redistribute

命令：redistribute { static | ospf | bgp} [metric <value>]。

　　　　no redistribute { static | ospf | bgp }。

功能：在 RIP 路由中引入其他路由协议的路由；本命令的 no 操作为取消引入。

参数：static 指定引入静态路由；ospf 指定引入 OSPF 路由；bgp 指定引入 BGP 路由；<value>指定以多大的路由权值引入路由，取值范围 1~16。

默认情况：RIP默认不引入其他路由。如果引入其他的路由协议而不指定它的metric值，则按默认路由权值default-metric引入。

命令模式：RIP 配置模式。

使用指南：采用本命令可以引入其他的路由作为 RIP 自己的路由，提高 RIP 的性能。

举例：在 RIP 报文中引入 OSPF 协议路由的路由权值为 5，静态路由的路由权值为 8。

Switch(Config-Router-Rip)#redistribute ospf metric 5
Switch(Config-Router-Rip)#redistribute static metric 8

rip broadcast

命令：rip broadcast。

　　　　no rip broadcast。

功能：配置三层交换机的的所有接口发 RIP 广播包或组播包；本命令的 no 操作禁止各端口发广播包或组播包，而只能在配置了 neighbor 的三层交换机之间发送 RIP 数据包。

默认情况：系统默认发 RIP 广播包。

命令模式：RIP 配置模式。

rip checkzero

命令：rip checkzero。

　　　　no rip checkzero。

功能：使用本命令对 RIPI 报文的零域进行检查；本命令的 no 操作为取消对零域的查零操作。由于 RIPII 的报文中没有零域，所以本命令对 RIPII 没有作用。

默认情况：系统默认对 RIPI 报文进行查零操作。

命令模式：RIP 协议配置模式。

使用指南：在 RIPI 的报文中必须有零域(zero field)，可以使用本命令来启动和禁止对 RIPI 报文的查零操作。如果进行查零操作时，若收到零域不为零的 RIPI 报文，系统将丢弃该 RIPI 的报文。

举例：设置不再对 RIPI 报文进行查零操作。

Switch(config-router-rip)#no ip checkzero

rip preference
命令：rip preference <value>。
　　　no rip preference。
功能：指定 RIP 协议的路由优先级；本命令的 no 操作为恢复默认值。
参数：<value>指定优先级的值，取值范围为 0~255。
默认情况：系统默认指定 RIP 的优先级是 120。
命令模式：RIP 协议配置模式。
使用指南：每一种路由协议都有自己的优先级，它的默认取值由具体的路由策略决定。优先级的高低将决定在核心路由表中的路由采取哪种路由算法获取的最佳路由。可以利用本命令手动调整 RIP 的优先级，优先级调整后，对新路由生效。由 RIP 协议的性质决定，RIP 的优先级不宜过高。
举例：设置 RIP 的优先级为 10。

Switch(config-router-rip)#rip prefenrence 10

router rip
命令：router rip。
　　　no router rip。
功能：开启 RIP 路由进程并进入 RIP 配置模式；本命令的 no 操作为关闭 RIP 路由协议。
默认情况：不运行 RIP 路由。
命令模式：全局配置模式。
使用指南：本命令是 RIP 路由协议的启动开关，进行 RIP 协议的其他配置要先打开本命令。
举例：启动 RIP 协议配置模式。

Switch(Config)#router rip
Switch(Config-Router-Rip)#

timer basic
命令：timer basic <update> <invalid> <holddown>。
　　　no timer basic。
功能：调整 RIP 计时器更新、期满、抑制的时间；本命令的 no 操作为恢复各项参数的默认值。
参数：<update>发送更新报文的时间间隔,单位秒,取值范围是 1~2147483647；<invalid>宣布 RIP 路由无效的时间段,单位秒,取值范围是 1~2147483647；<holddown>为宣布某路由无效后仍可在路由表中存在的时间段,单位秒,取值范围是 1~2147483647。

默认情况：<update>默认值是 30；<invalid>默认值是 180；<holddown>默认值是 120。

命令模式：RIP 协议配置模式。

使用指南：默认情况下，系统每 30 秒会广播 RIP 更新报文；当过了 180 秒不能收到某路由的更新报文，就认为该路由无效；但该路由还能在路由表中存在 120 秒，120 秒后，在路由表删除该路由。在调整 RIP 各项计时器时间时要注意，宣布 RIP 路由无效时间至少应大于 RIP 更新的时间，holddown 的时间段(即宣布 RIP 路由无效后，在路由表中删除该项路由的时间)也至少应该大于 RIP 更新的时间，且必须为整数倍。

举例：设置 RIP 路由表更新时间为 20 秒，宣布无效时间为 80 秒，删除路由项的时间为 60 秒。

Switch(Config-Router-Rip)#timer basic 20 80 60

version

命令：version {1| 2}。

　　　　no version。

功能：设置所有路由器接口发送/接收 RIP 数据报的版本；no 操作恢复默认设置。

参数：1 为 rip 版本 1；2 为 rip 版本 2。

默认情况：发送版本 1，接收版本 1 和 2 的数据报。

命令模式：RIP 协议配置模式。

使用指南：表示三层交换机各接口只发送/接收 RIP-I 数据报，2 表示三层交换机各接口只发送/接收 RIP-II 数据报。默认情况下，发送 RIP-II 数据报的同时接收 RIP-I 和 RIP-II 数据报。

举例：设置该交换机接口发送/接收 RIP 数据报的版本为 2。

Switch(config-router-rip)#version 2

相关命令：ip rip receive version　　　　ip rip send version。

show ip protocols

命令：show ip protocols。

功能：显示三层交换机当前运行的路由协议的信息。

命令模式：特权用户配置模式。

使用指南：根据本命令的输出信息，用户可以确认配置的路由协议是否正确和进行路由故障诊断。

举例：

Switch#sh ip protocols
RIP information
rip is turning on
default metrict 16
neighbour is:NULL
preference is 100

rip version information is:

interface	send version	receive version
vlan2	V2BC	V12
vlan3	V2BC	V12
vlan4	V2BC	V12(非粗体)

show ip route

使用 show ip route 命令可以显示路由表中关于 RIP 路由的目的 IP 地址、网络掩码以及下一跳 IP 地址或转发接口等信息。

例如，显示信息为：

Switch#show ip route
Total route items is 2, the matched route items is 2
Codes: C - connected, S - static, R - RIP derived, O - OSPF derived
         A - OSPF ASE, B - BGP derived, D - DVMRP derived

Destination	Mask	Nexthop	Interface	Pref
C 2.2.2.0	255.255.255.0	0.0.0.0	vlan1	0
R 7.7.7.0	255.255.255.0	2.2.2.8	vlan2	100

其中，R 表示 RIP 路由，即目的网络地址为 7.7.7.0、网络掩码为 255.255.255.0、下一跳地址为 2.2.2.8 以及转发接口为以太网口 vlan2 的一条路由为 RIP 路由，它的优先级为 100。

show ip protocols

使用 show ip protocols 命令可以显示当前三层交换机运行路由协议的信息。

例如，显示信息为：

Switch#sh ip protocols
RIP information:
Automatic network summarization is not in effect.
default metric for redistribute is :16
neigbour is:NULL
preference is :100
RIP version information is:

interface	send version	receive version
vlan1	V2BC	V12
vlan2	V2BC	V12

实训六　多层交换机的动态 rip 路由配置

一、实训设备

(1) DCS-3926S 交换机　　2 台。

(2) DCRS-5526S 交换机   2 台。

(3) PC 机   4 台。

(4) 直通网线   7 根。

二、实训拓扑

该实验拓扑结构如图 1-48 所示。

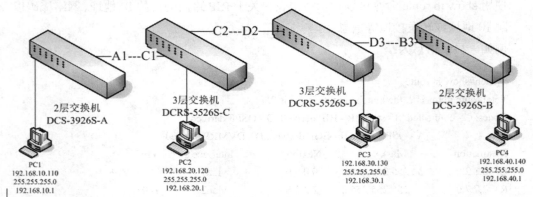

图 1-48 多层交换机的静态路由实验拓扑图

三、实训要求

按如图 1-46 所示进行连线。

实训结果：PC1---ping---PC2----PC3----PC4   全通。

四、实训步骤

交换机 A、B、C、D 做如下配置：

1. A 划分 vl 10(9-12)，vl 20(13-16)，Trunk (1)；

   B 划分 vl 30(17-20)，vl 40(21-24)，Trunk (3)；

   C 划分 vl 10(9-12)，vl 20(13-16)，vl 100(2)，Trunk (1)；

   D 划分 vl 30(17-20)，vl 40(21-24)，vl 101(2)，Trunk (3)。

2. 设置 3 层交换机 C 启用三层路由功能。

设置 vl 10、vl 20、vl 100 的接口地址：

int vl 10   192.168.10.1/24；
int vl 20   192.168.20.1/24；
int vl 100 192.168.100.1/30

设置 3 层交换机 D 启用三层路由功能；
设置 vl 30、vl 40、vl 101 的接口地址：

int vl 30   192.168.30.1/24；
int vl 40   192.168.40.1/24；

int vl 101 192.168.100.2/30

3. 方法 1：
配置 C 的动态路由 rip

C(config)#router rip
C(config-Router)#version 2
C(config)#int vl 10
C(config-if-vlan10)#ip rip work
C(config)#int vl 20
C(config-if-vlan20)#ip rip work
C(config)#int vl 100
C(config-if-vlan100)#ip rip work

配置 D 的动态路由 rip

D(config)#router rip
D(config-Router)#version 2
D(config)#int vl 30
D(config-if-vlan30)#ip rip work
D(config)#int vl 40
D(config-if-vlan40)#ip rip work
D(config)#int vl 101
D(config-if-vlan101)#ip rip work

验证：show ip rip   show ip route

方法 2：
配置 C 的动态路由 rip

C(config)#router rip
C(config-Router)#version 2
C(config-Router)#network vl 10
C(config-Router)#network vl 20
C(config-Router)#network vl 100

配置 D 的动态路由 rip

D(config)#router rip
D(config-Router)#version 2
D(config-Router)#network vl 30
D(config-Router)#network vl 40
D(config-Router)#network vl 101

验证：show ip rip   show ip route。

4. 根据连线位置正确配置 PC1、PC2、PC3、PC4 的地址(注意配上网关地址以及与 vlan 的对应关系)。

实验结果：PC1---ping---PC2----PC3----PC4   全通。

查看交换机状态：show vlan   show run   show ip route   show ip rip。

实验十七 三层交换机 OSPF 动态路由

一、实验目的

1. 掌握三层交换机之间通过 OSPF 协议实现网段互通的配置方法。
2. 理解 RIP 协议和 OSPF 协议内部实现的不同点。

二、应用环境

当两台三层交换机级联时，为了保证每台交换机上所连接的网段可以和另一台交换机上连接的网段互相通信，使用 OSPF 协议可以动态学习路由。

三、实验设备

1. DCRS-7604(或 6804)交换机   1 台。
2. DCRS-5526S 交换机   1 台。
3. PC 机   2~4 台。
4. Console 线   1~2 根。
5. 直通网线   2~4 根。

四、实验拓扑

该实验拓扑结构如图 1-49 所示。

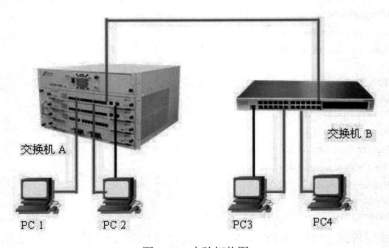

图 1-49 实验拓扑图

五、实验要求

1. 如表 1-52 所示，在交换机 A 和交换机 B 上分别划分基于端口的 VLAN。

表 1-52 交换机 A/B 的 VLAN 划分

交换机	VLAN	端口成员
交换机 A	10	1~8
	20	9~16
	100	24
交换机 B	30	1~8
	40	9~16
	101	24

2. 交换机 A 和 B 通过的 24 口级联。

3. 如表 1-53 所示，配置交换机 A 和 B 各 VLAN 虚拟接口的 IP 地址。

表 1-53 交换机 A/B 的 IP 配置

VLAN10	VLAN20	VLAN30	VLAN40	VLAN100	VLAN101
192.168.10.1	192.168.20.1	192.168.30.1	192.168.40.1	192.168.100.1	192.168.100.2

4. 按如表 1-54 所示进行 PC 的配置。

表 1-54 PC 的配置

设备	IP 地址	gateway	Mask
PC1	192.168.10.101	192.168.10.1	255.255.255.0
PC2	192.168.20.101	192.168.20.1	255.255.255.0
PC3	192.168.30.101	192.168.30.1	255.255.255.0
PC4	192.168.40.101	192.168.40.1	255.255.255.0

5. 验证：

没有 OSPF 路由协议之前：

PC1 与 PC2，PC3 与 PC4 可以互通。

PC1、PC2 与 PC3、PC4 不通。

配置 RIP 路由协议之后：

4 台 PC 之间都可以互通。

若实验结果和理论相符，则本实验完成。

六、实验步骤

第一步：交换机全部恢复出厂设置，配置交换机的 VLAN 信息(同实验 27)。

交换机 A：

DCRS-7604#conf

```
DCRS-7604(Config)#vlan 10
DCRS-7604(Config-Vlan10)#switchport interface ethernet 1/1-8
Set the port Ethernet1/1 access vlan 10 successfully
Set the port Ethernet1/2 access vlan 10 successfully
Set the port Ethernet1/3 access vlan 10 successfully
Set the port Ethernet1/4 access vlan 10 successfully
Set the port Ethernet1/5 access vlan 10 successfully
Set the port Ethernet1/6 access vlan 10 successfully
Set the port Ethernet1/7 access vlan 10 successfully
Set the port Ethernet1/8 access vlan 10 successfully
DCRS-7604(Config-Vlan10)#exit
DCRS-7604(Config)#vlan 20
DCRS-7604(Config-Vlan20)#switchport interface ethernet 1/9-16
Set the port Ethernet1/9 access vlan 20 successfully
Set the port Ethernet1/10 access vlan 20 successfully
Set the port Ethernet1/11 access vlan 20 successfully
Set the port Ethernet1/12 access vlan 20 successfully
Set the port Ethernet1/13 access vlan 20 successfully
Set the port Ethernet1/14 access vlan 20 successfully
Set the port Ethernet1/15 access vlan 20 successfully
Set the port Ethernet1/16 access vlan 20 successfully
DCRS-7604(Config-Vlan20)#exit
DCRS-7604(Config)#vlan 100
DCRS-7604(Config-Vlan100)#switchport interface ethernet 1/24
Set the port Ethernet1/24 access vlan 100 successfully
DCRS-7604(Config-Vlan100)#exit
DCRS-7604(Config)#
```

验证配置：

```
DCRS-7604#show vlan
VLAN Name                   Type         Media         Ports
--------------------------- ------------ ------------- ----------------------
1    default                Static       ENET          Ethernet1/17    Ethernet1/18
                                                       Ethernet1/19    Ethernet1/20
                                                       Ethernet1/21    Ethernet1/22
                                                       Ethernet1/23    Ethernet1/24
                                                       Ethernet1/25    Ethernet1/26
                                                       Ethernet1/27
10   VLAN0010               Static       ENET          Ethernet1/1     Ethernet1/2
                                                       Ethernet1/3     Ethernet1/4
                                                       Ethernet1/5     Ethernet1/6
                                                       Ethernet1/7     Ethernet1/8
20   VLAN0020               Static       ENET          Ethernet1/9     Ethernet1/10
                                                       Ethernet1/11    Ethernet1/12
```

				Ethernet1/13	Ethernet1/14
				Ethernet1/15	Ethernet1/16
100	VLAN0100	Static	ENET	Ethernet1/24	

DCRS-7604#

交换机 B：

DCRS-5526S(Config)#vlan 30
DCRS-5526S(Config-Vlan30)#switchport interface ethernet 0/0/1-8
Set the port Ethernet0/0/1 access vlan 30 successfully
Set the port Ethernet0/0/2 access vlan 30 successfully
Set the port Ethernet0/0/3 access vlan 30 successfully
Set the port Ethernet0/0/4 access vlan 30 successfully
Set the port Ethernet0/0/5 access vlan 30 successfully
Set the port Ethernet0/0/6 access vlan 30 successfully
Set the port Ethernet0/0/7 access vlan 30 successfully
Set the port Ethernet0/0/8 access vlan 30 successfully
DCRS-5526S(Config-Vlan30)#exit
DCRS-5526S(Config)#vlan 40
DCRS-5526S(Config-Vlan40)#switchport interface ethernet 0/0/9-16
Set the port Ethernet0/0/9 access vlan 40 successfully
Set the port Ethernet0/0/10 access vlan 40 successfully
Set the port Ethernet0/0/11 access vlan 40 successfully
Set the port Ethernet0/0/12 access vlan 40 successfully
Set the port Ethernet0/0/13 access vlan 40 successfully
Set the port Ethernet0/0/14 access vlan 40 successfully
Set the port Ethernet0/0/15 access vlan 40 successfully
Set the port Ethernet0/0/16 access vlan 40 successfully
DCRS-5526S(Config-Vlan40)#exit
DCRS-5526S(Config)#vlan 101
DCRS-5526S(Config-Vlan101)#switchport interface ethernet 0/0/24
Set the port Ethernet0/0/24 access vlan 101 successfully
DCRS-5526S(Config-Vlan101)#exit
DCRS-5526S(Config)#

验证配置：

DCRS-5526S#show vlan

VLAN	Name	Type	Media	Ports	
1	default	Static	ENET	Ethernet0/0/17	Ethernet0/0/18
				Ethernet0/0/19	Ethernet0/0/20
				Ethernet0/0/21	Ethernet0/0/22
				Ethernet0/0/23	
30	VLAN0030	Static	ENET	Ethernet0/0/1	Ethernet0/0/2
				Ethernet0/0/3	Ethernet0/0/4

				Ethernet0/0/5	Ethernet0/0/6
				Ethernet0/0/7	Ethernet0/0/8
40	VLAN0040	Static	ENET	Ethernet0/0/9	Ethernet0/0/10
				Ethernet0/0/11	Ethernet0/0/12
				Ethernet0/0/13	Ethernet0/0/14
				Ethernet0/0/15	Ethernet0/0/16
101	VLAN0101	Static	ENET	Ethernet0/0/24	

DCRS-5526S#

第二步：配置交换机各 vlan 虚接口的 IP 地址。

交换机 A：

DCRS-7604(Config)#int vlan 10

DCRS-7604(Config-If-Vlan10)#ip address 192.168.10.1 255.255.255.0

DCRS-7604(Config-If-Vlan10)#no shut

DCRS-7604(Config-If-Vlan10)#exit

DCRS-7604(Config)#int vlan 20

DCRS-7604(Config-If-Vlan20)#ip address 192.168.20.1 255.255.255.0

DCRS-7604(Config-If-Vlan20)#no shut

DCRS-7604(Config-If-Vlan20)#exit

DCRS-7604(Config)#int vlan 100

DCRS-7604(Config-If-Vlan100)#ip address 192.168.100.1 255.255.255.0

DCRS-7604(Config-If-Vlan100)#no shut

DCRS-7604(Config-If-Vlan100)#

DCRS-7604(Config-If-Vlan100)#exit

DCRS-7604(Config)#

交换机 B：

DCRS-5526S(Config)#int vlan 30

DCRS-5526S(Config-If-Vlan30)#ip address 192.168.30.1 255.255.255.0

DCRS-5526S(Config-If-Vlan30)#no shut

DCRS-5526S(Config-If-Vlan30)#exit

DCRS-5526S(Config)#interface vlan 40

DCRS-5526S(Config-If-Vlan40)#ip address 192.168.40.1 255.255.255.0

DCRS-5526S(Config-If-Vlan40)#exit

DCRS-5526S(Config)#int vlan 101

DCRS-5526S(Config-If-Vlan101)#ip address 192.168.100.2 255.255.255.0

DCRS-5526S(Config-If-Vlan101)#exit

DCRS-5526S(Config)#

第三步：按如表 1-55 所示配置各 PC 的 IP 地址，注意配置网关。

表 1-55 PC 的配置

设备	IP 地址	gateway	Mask
PC1	192.168.10.101	192.168.10.1	255.255.255.0
PC2	192.168.20.101	192.168.20.1	255.255.255.0
PC3	192.168.30.101	192.168.30.1	255.255.255.0
PC4	192.168.40.101	192.168.40.1	255.255.255.0

第四步：如表 1-56 所示，验证 PC 之间的连通性。

表 1-56 验证 PC 之间的连通性

PC	端口	PC	端口	结果	原因
PC1	A：1/1	PC2	A：1/9	通	
PC1	A：1/1	Vlan 100	A：1/24	通	
PC1	A：1/1	Vlan 101	B：0/0/24	不通	
PC1	A：1/1	PC3	B：0/0/1	不通	

查看路由表，进一步分析上一步的现象原因。

交换机 A：

DCRS-7604#show ip route

Total route items is 3, the matched route items is 3

Codes: C - connected, S - static, R - RIP derived, O - OSPF derived

      A - OSPF ASE, B - BGP derived, D - DVMRP derived

Destination	Mask	Nexthop	Interface	Preference
C 192.168.10.0	255.255.255.0	0.0.0.0	Vlan10	0
C 192.168.20.0	255.255.255.0	0.0.0.0	Vlan20	0
C 192.168.100.0	255.255.255.0	0.0.0.0	Vlan100	0

DCRS-7604#

交换机 B：

DCRS-5526S#show ip route

Total route items is 3, the matched route items is 3

Codes: C - connected, S - static, R - RIP derived, O - OSPF derived

      A - OSPF ASE, B - BGP derived, D - DVMRP derived

Destination	Mask	Nexthop	Interface	Preference
C 192.168.30.0	255.255.255.0	0.0.0.0	Vlan30	0
C 192.168.40.0	255.255.255.0	0.0.0.0	Vlan40	0
C 192.168.100.0	255.255.255.0	0.0.0.0	Vlan101	0

DCRS-5526S#

第五步：启动 OSPF 协议，并将对应的直连网段配置到 OSPF 进程中。

交换机 A：

DCRS-7604(Config)#router ospf

```
OSPF protocol is working, please waiting.......
OSPF protocol has enabled!
DCRS-7604(Config-Router-Ospf)#exit
DCRS-7604(Config)#interface vlan 10
DCRS-7604(Config-If-Vlan10)#ip ospf enable area 0
DCRS-7604(Config-If-Vlan10)#
DCRS-7604(Config)#interface vlan 20
DCRS-7604(Config-If-Vlan20)#ip ospf enable area 0
DCRS-7604(Config-If-Vlan20)#exit
DCRS-7604(Config)#interface vlan 100
DCRS-7604(Config-If-Vlan100)#
DCRS-7604(Config-If-Vlan100)#ip ospf enable area 0
DCRS-7604(Config-If-Vlan100)#exit
DCRS-7604(Config)#
```

验证配置：

```
DCRS-7604#show ip route
Total route items is 4, the matched route items is 4
Codes: C - connected, S - static, R - RIP derived, O - OSPF derived
       A - OSPF ASE, B - BGP derived, D - DVMRP derived
Destination      Mask            Nexthop         Interface    Preference
C  192.168.10.0  255.255.255.0   0.0.0.0         Vlan10       0
O  192.168.30.0  255.255.255.0   192.168.100.2   Vlan100      110
O  192.168.40.0  255.255.255.0   192.168.100.2   Vlan100      110
C  192.168.100.0 255.255.255.0   0.0.0.0         Vlan100      0
DCRS-7604#
```

交换机 B：(5526S 的配置界面和 7604 是一样的)。

第六步：如表 1-57 所示，验证 PC 之间的连通。

表 1-57 验证 PC 之间的连通性

PC	端口	PC	端口	结果	原因
PC1	A：1/1	PC2	A：1/9	通	
PC1	A：1/1	Vlan 100	A：1/24	通	
PC1	A：1/1	Vlan 101	B：0/0/24	通	
PC1	A：1/1	PC3	B：0/0/1	通	

七、注意事项和排错

在配置、使用 OSPF 协议时，可能会由于物理连接、配置错误等原因导致 OSPF 协议未能正常运行。因此，用户应注意以下要点：

首先应该保证物理连接的正确无误。

其次，保证接口和链路协议是 UP(使用 show interface 命令)。

然后在各接口上配置不同网段的 IP 地址。

最后，先启动 OSPF 协议(使用 router ospf 命令)再在相应接口配置所属 OSPF 域；接着，注意 OSPF 协议的自身特点——OSPF 骨干域(0 域)必须保证是连续的，如果不连续使用虚连接(virtual link)来保证，所有非 0 域只能通过 0 域与其他非 0 域相连，不允许非 0 域直接相连；边界三层交换机是指该三层交换机的一部分接口属于 0 域，而另外一部分接口属于非 0 域；对于广播网等多路访问网，需要选举指定三层交换机 DR。

八、思考题

本实验只体现了 area0 的配置方法，大家可以参考理论教材和用户手册尝试 OSPF 运行在多区域的配置方法。

OSPF 的配置命令

```
default redistribute cost
default redistribute interval
default redistribute limit
default redistribute tag
default redistribute type
ip opsf authentication
ip ospf cost
ip opsf dead-interval
ip ospf enable area
ip ospf hello-interval
ip ospf passive-interface
ip ospf priority
ip ospf retransmit-interval
ip ospf transmit-delay
network
preference
redistribute ospfase
router id
router ospf
stub cost
virtuallink neighborid
show ip ospf
show ip ospfase
show ip ospf cumulative
show ip ospf database
show ip ospf interface
show ip ospf neighbor
show ip ospf routing
show ip ospf virtual-links
show ip protocols
debug ip ospf event
```

debug ip ospf lsa
debug ip ospf packet
debug ip ospf spf

default redistribute cost

命令：default redistribute cost <cost>。

　　　　no default redistribute cost。

功能：配置 OSPF 引入外部路由时默认的花费值；本命令的 no 操作为恢复默认值。

参数：<cost>为花费值，取值范围是 1~65535。

默认情况：默认设置引入的花费值为 1。

命令模式：OSPF 协议配置模式。

使用指南：OSPF 路由协议引入由其他路由协议发现的路由时，把这些路由信息作为自己的自治系统外部的路由信息。引入外部路由信息需要一些额外的参数，如：路由的默认花费和默认的标记等。本命令提供给用户可据实际情况设置合理的引入外部路由时默认花费值。

举例：设置 OSPF 引入外部路由的默认花费值为 20。

Switch(Config-Router-Ospf)#default redistribute cost 20

default redistribute interval

命令：default redistribute interval <time>。

　　　　no default redistribute interval。

功能：配置 OSPF 引入外部路由的时间间隔；本命令的 no 操作为恢复默认值。

参数：<time>为引入外部路由的时间间隔，单位为秒，取值范围为 1~65535。

默认情况：OSPF 引入外部路由的时间间隔默认为 1 秒。

命令模式：OSPF 协议配置模式。

使用指南：OSPF 会定期引入外部的路由信息，并将这些路由信息传播到整个自治系统中，本命令用于修改引入外部路由信息的时间间隔。

举例：OSPF 引入外部路由的时间间隔为 3 秒。

Switch(Config-Router-Ospf)#default redistribute interval 3

default redistribute limit

命令：default redistribute limit <routes>。

　　　　no default redistribute limit。

功能：配置 OSPF 一次可引入外部路由的最大值；本命令的 no 操作为恢复默认值。

参数：<routes>为引入路由数量的最大值，取值范围是 1~65535。

默认情况：OSPF 引入外部路由数量的最大值默认为 100。

命令模式：OSPF 协议配置模式。

使用指南：OSPF 定期引入外部的路由信息并将它们传播到整个自治系统中，本命令规定在一次能够引入的外部路由信息的最大条数。

举例：设置 OSPF 一次最多可引入 110 条外部路由。

Switch(Config-Router-Ospf)#default redistribute limit 110

default redistribute tag

命令：default redistribute tag <tag>。

　　　　no default redistribute tag。

功能：配置引入外部路由时默认的标记值；本命令的 no 操作为恢复默认值。

参数：<tag>为标记值，取值范围是 0~4294967295。

默认情况：默认值为 0。

命令模式：OSPF 协议配置模式。

使用指南：OSPF 路由协议引入由其他路由协议发现的路由时，把这些路由信息作为自己自治系统外部的路由信息。引入外部路由信息需要一些额外的参数，如：路由的默认花费和默认的标记等。本命令为用户提供路由标记标识协议相关的信息。

举例：设置 OSPF 引入外部路由时默认的标记值为 20000。

Switch(Config-Router-Ospf)#default redistribute tag 20000

default redistribute type

命令：default redistribute type { 1 | 2 }。

　　　　no default redistribute type。

功能：配置引入外部路由时默认的类型；本命令的 no 操作为恢复默认值。

参数：1 | 2 分别表示第一类外部路由和第二类外部路由。

默认情况：系统默认认为引入的外部路由为第二类外部路由。

命令模式：OSPF 协议配置模式。

使用指南：OSPF 在协议中规定了两类外部路由信息的代价选择方式：第一类外部路由和第二类外部路由。第一类外部路由的代价=外部路由的通告代价+从某个三层交换机到通告三层交换机(AS 外部三层交换机)的代价。第二类外部路由的代价=外部路由的通告代价。第一类和第二类外部路由同时存在的时，第一类外部路由代价的优先级高。

举例：设置 OSPF 引入外部路由时默认的类型为 type 1。

Switch(Config-Router-Ospf)#default redistribute type 1

ip ospf authentication

命令：ip ospf authentication { simple <auth_key>| md5 <auth_key> <key_id> }。

　　　　no ip ospf authentication。

功能：指定接口上接受 OSPF 报文所需要的验证方式；本命令的 no 操作为取消验证。

参数：simple 为简单验证方式；md5 为 MD5 加密验证方式；<auth_key>为验证密钥，为连续的字符串，简单验证方式下最大长度为 8 字节，MD5 验证方式下最大长度为 16 字节；<key_id>为 MD5 验证方式时的验证字，取值范围是 1~255。

默认情况：接口上接受 OSPF 报文默认不需要验证。

命令模式：接口配置模式。

使用指南：密钥的值将写入 OSPF 报文中，为保证三层交换机与相邻三层交换机之间 OSPF 报文的正常收发，必须在对端设置相同的密钥参数。

举例：在 OSPF 接口 vlan1 配置 MD5 验证方式，验证密码为 123abc。

Switch(Config-If-Vlan1)#ip ospf authentication md5 123abc 1

ip ospf cost

命令：ip ospf cost <cost>。

　　　no ip ospf cost。

功能：指定接口运行 OSPF 协议所需的代价；本命令的 no 操作为恢复默认值。

参数：<cost>为 OSPF 协议所需花费的值，取值范围是 1~65535。

默认情况：接口默认的 OSPF 协议所需花费的值为 1。

命令模式：接口配置模式。

举例：将接口 vlan1 的 OSPF 路由代价配置成 3。

Switch(Config-If-Vlan1)#ip ospf cost 3

ip ospf dead-interval

命令：ip ospf dead-interval <time>。

　　　no ip ospf dead-interval。

功能：指定相邻三层交换机路由失效的时间长度；本命令的 no 操作为恢复默认值。

参数：<time>为相邻三层交换机失效的时间长度，单位为秒，取值范围是 1~65535。

默认情况：三层交换机失效的时间长度默认为 40 秒（通常是 hello-interval 的 4 倍）。

命令模式：接口配置模式。

dead-interval 参数一致，且至少为 hello-interval 值的 4 倍。

举例：将接口 vlan1 的 OSPF 路由失效时间设置为 80s。

Switch(Config-If-Vlan1)#ip ospf dead-interval 80

ospf enable area

命令：ip ospf enable area <area_id>。

　　　no ip ospf enable area。

功能：配置接口属于某个 OSPF 区域；本命令的 no 操作为取消该配置。

参数：<area_id>为该接口所属区域的区域号，取值范围是 0~4294967295。

默认情况：接口默认没有被配置成属于某个区域。

命令模式：接口配置模式。

使用指南：要在某一个接口上运行 OSPF 协议，必须首先指定该接口属于一个区域。

举例：将接口 vlan1 配置为属于 1 域。

Switch(Config-If-Vlan1)#ip ospf enable area 1

ip ospf hello-interval

命令：ip ospf hello-interval <time>。

no ip ospf hello-interval。

功能：指定在接口上发送 HELLO 报文的时间间隔；本命令的 no 操作为恢复默认值。

参数：<time>为发送 HELLO 报文的时间间隔，单位为秒，取值范围是 1~255。

默认情况：接口默认发送 HELLO 报文的间隔时间为 10 秒。

命令模式：接口配置模式。

使用指南：HELLO数据包是一种最常见的一种数据包，它周期性地被发送至邻接三层交换机，用于发现和维持邻接关系、选举DR与BDR。用户设置的hello-interval的值将写入HELLO报文中，并随HELLO报文传送。hello-interval的值越小，则网络拓扑结构的变化将被越快发现，同时路由开销也增加。为了使OSPF协议的正常运行，必须保证和该接口相邻的三层交换机之间的hello-interval参数一致。

举例：配置接口 vlan1 发送 HELLO 报文的间隔时间为 20 秒。

Switch(Config-If-Vlan1)#ip ospf hello-interval 20

相关命令：ip ospf dead-interval。

ip ospf passive-interface

命令：ip ospf passive-interface。

no ip ospf passive-interface。

功能：将接口设置为只收不发 OSPF 报文；本命令的 no 操作为取消该项配置。

默认情况：接口默认状态是收发 OSPF 报文。

命令模式：接口配置模式。

举例：配置以太网口接口 vlan1 只收不发 OSPF 报文。

Switch(Config-If-Vlan1)#ip ospf passive-interface

ip ospf priority

命令：ip ospf priority <priority>。

no ip ospf priority。

功能：配置接口在选举"指定三层交换机"(DR)时的优先级；本命令 no 操作为恢复默认值。

参数：<priority>为优先级，合法的取值范围是 0~255。

默认情况：接口在选举指定三层交换机时默认的优先级值为 1。

命令模式：接口配置模式。

使用指南：当连在同一网段的两台三层交换机都想成为"指定三层交换机"时，根据优先级的值来决定谁是"指定三层交换机"，通常选择优先级高的作为"指定三层交换机"；

如果优先级值相等，则选 router-id 号大的。当一台三层交换机的优先级值为 0 时，这台三层交换机将不会被选举为"指定三层交换机"或"备份指定三层交换机"。

举例：配置接口选举指定三层交换机 DR 中的优先级。将接口 vlan1 配置成没有选举权利，即 priority 值为 0。

Switch(Config-If-Vlan1)#ip ospf priority 0

ip ospf retransmit-interval

命令：ip ospf retransmit-interval <time>。

　　　　no ip ospf retransmit-interval。

功能：指定接口与邻接三层交换机之间传送链路状态宣告(LSA)时的重传间隔；本命令的 no 操作为恢复默认值。

参数：<time>为与邻接三层交换机之间传送链路状态宣告时的重传间隔，单位为秒，取值范围是 1~65535。

默认情况：默认重传间隔为 5 秒。

命令模式：接口配置模式。

使用指南：当一台三层交换机向它的邻居传送链路状态宣告时，它将保持链路状态宣告直至收到对方的确认，若在时间间隔内没有收到确认报文，则三层交换机将重传链路状态宣告。重传间隔的值必须大于两台三层交换机传送报文一个来回的时间。

举例：设置接口 vlan1 重传 lsa 的时间为 10 秒。

Switch(Config-If-Vlan1)#ip ospf retransmit 10

ip ospf transmit-delay

命令：ip ospf tranmsit-delay <time>。

　　　　no ip ospf transmit-delay。

功能：设置在接口上传送链路状态宣告(LSA)的时延值；本命令的 no 操作为恢复默认值。

参数：<time>为接口上传送链路状态宣告的时延值，单位为秒，取值范围是 1~65535。

默认情况：接口上传送链路状态宣告的默认时延值为 1 秒。

命令模式：接口配置模式。

使用指南：链路状态宣告在本三层交换机中会随时间老化，但在网络传输过程中却不会，因此在发送链路状态宣告之前增加 transmit-delay 的时延，使之能在老化之前将链路状态宣告发送出去。

举例：设置接口 vlan1 发送 LSA 的时延为 2 秒。

Switch(Config-If-Vlan1)#ip ospf transmit-delay 2

network

命令：network <network> <mask> area <area_id> [advertise | notadvertise]。

　　　　no network <network> <mask> area <area_id>。

功能：为三层交换机的各个网络定义所属区域；本命令的 no 操作为删除该项配置。

参数：<network>和<mask>为网络 IP 地址和地址通配符位，点分十进制格式；<area_id>为区域号，取值范围是 0~4294967295；advertise | notadvertise 指定是否将到这一网络范围路由的摘要信息广播出去。

默认情况：系统默认没有配置网络所属的区域；若配置了，则默认认为是广播摘要信息。

命令模式：OSPF 协议配置模式。

使用指南：一旦将某一网络的范围加入到区域中，所有该网络的内部路由都不再被独立地广播到别的区域，而只是广播整个网络范围路由的摘要信息。引入网络范围和对该范围的限定，可以减少区域间路由信息的交流量。

举例：定义网络范围 10.1.1.0 255.255.255.0 并加入到区域 1 中。

Switch(Config-Router-Ospf)#network 10.1.1.0 255.255.255.0 area 1

preference

命令：preference [ase] <preference>。

　　　no preference [ase]。

功能：配置 OSPF 协议在各路由协议之间的优先级，以及引入的自治系统外部路由的优先级；本命令的 no 操作为恢复默认值。

参数：ase 表示指定引入自治系统外部路由的优先级；<preference>为优先级值，取值范围是 1~255。

默认情况：OSPF 协议的默认优先级为 10；引入的外部路由协议的默认优先级为 150。

命令模式：OSPF 协议配置模式。

使用指南：由于三层交换机上可能同时运行多个动态路由协议，就存在各个路由协议之间路由信息共享和选择的问题。所以为每一种路由协议指定了一个默认的优先级，当不同协议发现同一条路由时，优先级高的协议将起决定作用。优先级更改后对新构造的路由开始有效。由 OSPF 的性质决定，OSPF 的优先级不宜过低。

举例：设置 OSPF 引入 ase 路由时的优先级为 20。

Switch(Config- Router-Ospf)#preference ase 20

redistribute ospfase

命令：redistribute ospfase { bgp |connected | static | rip} [type { 1 | 2 }] [tag <tag>] [metric<cost_value>]。

　　　no redistribute ospfase { bgp |connected | static | rip}。

功能：引入 bgp 路由、直连路由、静态路由和 RIP 路由作为外部路由信息；本命令的 no 操作为取消引入的外部路由信息。

参数：bgp 表示引入 BGP 路由作为外部路由信息；connected 表示引入直连路由作为外部路由信息；static 表示引入静态路由作为外部路由信息；rip 表示引入 RIP 协议发现路由

作为外部路由信息；type 指定外部路由类型，1 | 2 分别表示第一类外部路由和第二类外部路由；tag 指定路由的标记，<tag>为路由的标记值，取值范围为 0~4294967295；metric 指定路由的权值，<cost_value>为路由的权值，取值范围为 1~16777215。

默认情况：OSPF 默认不引入外部路由。

命令模式：OSPF 协议配置模式。

使用指南：三层交换机上各动态路由协议之间是可以互相共享路由信息的，由于 OSPF 的特性，其他的路由协议发现的路由总被当作自治系统外部的路由信息处理。

举例：在 OSPF 路由中引入 RIP 路由作为第一类外部路由，引入标记值为 3，引入代价为 20。

Switch(Config-Router-Ospf)#redistribute ospfase rip type 1 tag 3 metric 20

router id

命令：router id <router_id>。

　　no router id。

功能：配置运行 OSPF 协议三层交换机的 ID 号；本命令的 no 操作为取消三层交换机的 ID 号。

参数：<router_id>为三层交换机 ID 号，点分十进制格式。

默认情况：系统默认为不配置三层交换机 ID 号，协议运行时从各接口的 IP 地址中选择其中一个地址作为三层交换机 ID 号。

命令模式：全局配置模式。

使用指南：OSPF协议运行时把三层交换机的ID号作为本三层交换机在自治系统中的唯一标识，通常选取三层交换机中运行OSPF协议的某个接口的IP地址作为ID号。CRS-7604三层交换机默认使用该交换机的最先UP起来的三层接口的IP地址为router id。若三层交换机所有接口上都没有配置IP地址时，必须使用本命令指定三层交换机的ID号，否则OSPF协议无法运行。三层交换机ID号的变化在OSPF重启后才起作用。

举例：指定三层交换机的 ID 号为 10.1.120.1。

Switch(Config)#router id 10.1.120.1

router ospf

命令：router ospf。

　　no router ospf。

功能：启动 OSPF 协议，开启后进入 OSPF 模式；本命令的 no 操作为关闭 OSPF 协议。

默认情况：系统默认不运行 OSPF 协议。

命令模式：全局配置模式。

使用指南：使用本命令运行或终止 OSPF 协议。有关 OSPF 的配置，只有在系统运行了 OSPF 后才能生效。

举例：配置本交换机运行 OSPF。

Switch(Config)#router ospf

stub cost

命令：stub cost <cost> area <area_id>。
　　　no stub area <area_id>。

功能：将一个区域定义成 STUB 区域；本命令的 no 操作为取消该定义。

参数：<cost>为 STUB 区域默认路由的花费值，取值范围 1~65535；<area_id>为 STUB 区域的区域号，取值范围是 1~4294967295。

默认情况：系统默认没有配置 STUB 区域。

命令模式：OSPF 协议配置模式。

使用指南：当一个区域只有一个出口点时(只与一个三层交换机相连)，或不必为每个外部目的地选择出口点时，它就可以被配置成 STUB 域。在 STUB 区域中类型 4LSA(ASBR 汇总 LSA)和类型 5LSA(AS 外部 LSA)两种 LSA 不允许泛滥进入/通过，可以节省该区域内部各三层交换机处理外部路由信息所花费的资源。

举例：将 1 域配置成 STUB 域，默认路由的代价为 60。

Switch(Config-Router-Ospf)#stub cost 60 area 1

实训七 多层交换机之间的动态 ospf 配置

一、实训设备

(1) DCS-3926S 交换机　　2 台。
(2) DCRS-5526S 交换机　　2 台。
(3) PC 机　　4 台。
(4) 直通网线　　7 根。

二、实训拓扑

该实验拓扑结构如图 1-50 所示。

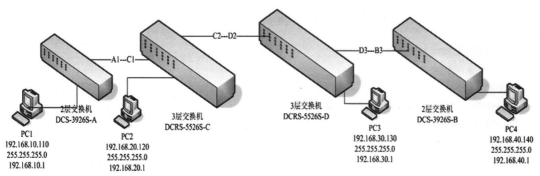

图 1-50　实验拓扑图

三、实训要求

如图1-49所示进行连线。

1. 交换机A、B、C、D做如下配置：

   A 划分 vl 10(9-12)，vl 20(13-16)，Trunk (1)；

   B 划分 vl 30(17-20)，vl 40(21-24)，Trunk (3)；

   C 划分 vl 10(9-12)，vl 20(13-16)，vl 100(2)，Trunk (1)；

   D 划分 vl 30(17-20)，vl 40(21-24)，vl 101(2), Trunk (3)。

2. 设置3层交换机C启用三层路由功能。

设置 vl 10、vl 20、vl 100 的接口地址：

int vl 10　　192.168.10.1/24；
int vl 20　　192.168.20.1/24；
int vl 100 192.168.100.1/30。

3. 设置3层交换机D启用三层路由功能。

设置 vl 30、vl 40、vl 101 的接口地址：

int vl 30　　192.168.30.1/24；
int vl 40　　192.168.40.1/24；
int vl 101 192.168.100.2/30。

4. 根据连线位置正确配置PC1、PC2、PC3、PC4的地址(注意配上网关地址以及与vlan的对应关系)。

实验结果：PC1---ping---PC2----PC3----PC4　　全通。

查看交换机状态：show vlan　　show run　　show ip route　　show ip rip。

四、实训步骤

主要配置方法参考：

1. 交换机一般配置：略。
2. 动态ospf路由协议启用方法：

方法1：启动交换机C的动态路由ospf协议

C(config)#router ospf
C(config-Router-ospf)#exit
C(config)#int vl 10
C(config-if-vlan10)#ip ospf enable area 0
C(config)#int vl 20
C(config-if-vlan20)#ip ospf enable area 0
C(config)#int vl 100
C(config-if-vlan100)#ip ospf enable area 0

验证：sh ip rip　　　　sh ip route

启动交换机 D 的动态路由 ospf 协议

D(config)#router ospf
D(config-Router-ospf)#exit
D(config)#int vl 30
D(config-if-vlan30)#ip ospf enable area 0
D(config)#int vl 40
D(config-if-vlan40)#ip ospf enable area 0
D(config)#int vl 101
D(config-if-vlan101)#ip ospf enable area 0

验证：sh ip rip   sh ip route。
因设备版本差别也可使用下列方法：
方法 2：启动交换机 C 的动态路由 ospf 协议

C(config)#router ospf
C(config-Router)#network 192.168.10.0 255.255.255.0 area 1
C(config-Router)#network 192.168.20.0 255.255.255.0 area 1
C(config-Router)#network 192.168.100.0 255.255.255.0 area 1

启动交换机 D 的动态路由 ospf 协议

C(config)#router ospf
C(config-Router)#network 192.168.30.0 255.255.255.0 area 1
C(config-Router)#network 192.168.40.0 255.255.255.0 area 1
C(config-Router)#network 192.168.100.0 255.255.255.0 area 1

验证：sh ip rip   sh ip route。

第二章 路由器实验

实验一 路由器接口简介

一、实验目的

1. 掌握路由器各接口的外观。
2. 接口的功能。
3. 接口的表示方法。

二、应用环境

路由器是三层设备,主要功能是进行路径选择和广域网的连接。与交换机相比,接口数量要少很多,但功能要强大得多,这些功能在外观上就是接口、模块的类型比较多,当然价格有很大差异,通常高端的设备都是模块化的,支持的模块类型也很丰富。

三、实验设备

DCR 1702　　一台。

四、实验拓扑

该实验拓扑结构如图 2-1 所示。

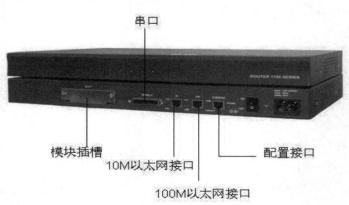

图 2-1　实验拓扑图

五、实验要求

1. 以太网接口有 100M 和 10M 之分。
2. 注意观察接口和模块上的标志。
3. 注意插槽上标号,从靠近电源开始,依次是 0、1、2、3。
4. 各接口也从 0 开始的。

六、实验步骤

1. 一个固定的 Console 接口,用于本地配置路由器或路由器的软件升级。
2. 分别有 10/100M 自适应的 RJ45 以太网接口,主要用来连接以太网。
3. DCR-1751 有一个模块扩展插槽,DCR-1750 有 3 个模块扩展插槽,如表 2-1 所示,可以选用表 2-1 中的各种模块(部分)。

表 2-1　模块说明表

序号	模块名称	模块描述	Slot 1	Slot 2	Slot 3
1	MR-WIC-1ETH	单路 10M 以太网接口卡	Yes	Yes	/
2	MR-WIC-2ETH	双路 10M 以太网接口卡	Yes	Yes	/
3	MR-WIC-1T	单路同步/异步串口卡	Yes	Yes	/
4	MR-WIC-2T	双路同步/异步串口卡	Yes	Yes	/
5	MR-WIC-1E1T	单路 10M 以太网+单同步/异步串口卡	Yes	Yes	/
6	MR-WIC-1CE1	单路 E1 接口卡	/	Yes	/
7	MR-WIC-1B-S/T	单路 ISDN BRI S/T 接口卡	/	Yes	/
8	MR-VIC-2FXS	双口 FXS 语音接口卡	Yes	Yes	Yes
9	MR-VIC-2FXO	双口 FXO 语音接口卡	Yes	Yes	Yes
10	MR-VIC-2E&M	双口 E&M 语音接口卡	Yes	Yes	Yes
11	MR-EIC-8A	8 路异步通信卡	Yes	Yes	Yes

七、注意事项和排错

1. 加装和拆卸模块一定要先关闭电源。
2. 串口不要带电插拔。

实验二　路由器的基本管理方法

一、实验目的

1. 掌握带外的管理方法:通过 Console 接口配置。
2. 掌握带内的管理方法:通过 Telnet 方式配置。
3. 掌握带内的管理方法:通过 Web 方式配置。

二、应用环境

1. 设备的初始配置一般都是通过 Console 接口进行。远程管理通常通过带内的方式。
2. 给相应的接口配置了 IP 地址，开启了相应的服务以后，才能进行带内的管理。

三、实验设备

1. DCR-1702   一台。
2. DCR-2611   一台。
3. PC 机   一台。
4. Console 线缆、网线   各一条。

四、实验拓扑

该实验拓扑结构如图 2-2 所示。

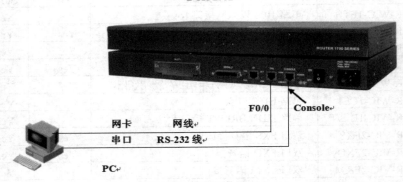

图 2-2   实验拓扑图

五、实验要求

设备接口配置如表 2-2 所示。

表 2-2   设备接口配置表

DCR-1702		PC 机	
Console		串口	
F0/0	192.168.2.1	网卡	192.168.2.2

六、实验步骤

带外管理方法：(本地管理)。

第一步：将配置线的一端与路由器的 Console 口相连，另一端与 PC 的串口相连，如上图 2-2 所示。

第二步：在 PC 上运行终端仿真程序。选择"开始"|"程序"|"附件"|"通讯"|"超

级终端"命令，打开如图 2-3 所示的"属性"对话框，设置终端的硬件参数(包括串口号)，波特率：9600 了；数据位：8；奇偶校验：无；停止位：1； 数据流控制：无。

图 2-3　端口配置

第三步：路由器加电，超级终端会显示路由器自检信息，自检结束后出现命令提示"Press RETURN to get started"。

```
System Bootstrap, Version 0.1.8
Serial num:8IRT01V11B01000054 ,ID num:000847
Copyright (c) 1996-2000 by China Digitalchina CO.LTD
DCR-1700 Processor MPC860T @ 50Mhz
The current time: 2067-9-12 6:31:30
Loading DCR-1702.bin......
Start Decompress DCR-1702.bin
####################################################################################
####################################################################################
####################################################################################
#######################
Decompress 3587414 byte,Please wait system up..
Digitalchina Internetwork Operating System Software
DCR-1700 Series Software , Version 1.3.2E, RELEASE SOFTWARE
System start up OK
Router console 0 is now available
Press RETURN to get started
```

第四步：按回车键进入用户配置模式。DCR-1702系列路由器出厂时没有定义密码，用户按回车键直接进入普通用户模式，可以使用权限允许范围内的命令，需要帮助可以随时输入"？"，输入enable，按回车键则进入超级用户模式。这时候用户拥有最大权限，可以任意配置，需要帮助可以随时输入"？"。

```
Router-A>enable                                          ! 进入特权模式
Router-A#2004-1-1 00:04:39 User DEFAULT enter privilege mode from console 0, level = 15
  Router-A#?                                             ! 查看可用的命令
```

Cd	-- Change directory
chinese	-- Help message in Chinese
chmem	-- Change memory of system
chram	-- Change memory
clear	-- Clear something
config	-- Enter configurative mode
connect	-- Open a outgoing connection
copy	-- Copy configuration or image data
date	-- Set system date
debug	-- Debugging functions
delete	-- Delete a file
dir	-- List files in flash memory
disconnect	-- Disconnect an existing outgoing network connection
download	-- Download with ZMODEM
enable	-- Turn on privileged commands
english	-- Help message in English
enter	-- Turn on privileged commands
exec-script	-- Execute a script on a port or line
exit	-- Exit / quit
format	-- Format file system
help	-- Description of the interactive help system
history	-- Look up history

Router-A#ch?   !使用?帮助

chinese	-- Help message in Chinese
chmem	-- Change memory of system
chram	-- Change memory

Router-A#chinese   !设置中文帮助
Router-A#?   !再次查看可用命令

cd	-- 改变当前目录
chinese	-- 中文帮助信息
chmem	-- 修改系统内存数据
chram	-- 修改内存数据
clear	-- 清除
config	-- 进入配置态
connect	-- 打开一个向外的连接
copy	-- 复制配置方案或内存映像
date	-- 设置系统时间
debug	-- 分析功能
delete	-- 删除一个文件
dir	-- 显示闪存中的文件
disconnect	-- 断开活跃的网络连接
download	-- 通过 ZMODEM 协议下载文件
enable	-- 进入特权方式
english	-- 英文帮助信息

enter	-- 进入特权方式
exec-script	-- 在指定端口运行指定的脚本
exit	-- 退回或退出
format	-- 格式化文件系统
help	-- 交互式帮助系统描述
history	-- 查看历史
keepalive	-- 保活探测

--More--

带内远程的管理方法：(Telnet 方式)。

第五步：设置路由器以太网接口地址并验证。

```
Router>enable                                           ！进入特权模式
Router #config                                          ！进入全局配置模式
Router-A_config#interface f0/0                          ！进入接口模式
Router-A_config_f0/0#ip address 192.168.2.1 255.255.255.0   ！设置 IP 地址
Router-A_config_f0/0#no shutdown
Router-A_config_f0/0#^Z
Router-A#show interface f0/0                            ！验证
FastEthernet0/0 is up, line protocol is up              ！接口和协议都必须 up
address is 00e0.0f18.1a70
    Interface address is 192.168.2.1/24
    MTU 1500 bytes, BW 100000 kbit, DLY 10 usec
    Encapsulation ARPA, loopback not set
    Keepalive not set
    ARP type: ARPA, ARP timeout 04:00:00
    60 second input rate 0 bits/sec, 0 packets/sec!
    60 second output rate 6 bits/sec, 0 packets/sec!
    Full-duplex, 100Mb/s, 100BaseTX, 1 Interrupt
        0 packets input, 0 bytes, 200 rx_freebuf
        Received 0 unicasts, 0 lowmark, 0 ri, 0 throttles
        0 input errors, 0 CRC, 0 framing, 0 overrun, 0 long
        1 packets output, 46 bytes, 50 tx_freebd, 0 underruns
        0 output errors, 0 collisions, 0 interface resets
        0 babbles, 0 late collisions, 0 deferred, 0 err600
        0 lost carrier, 0 no carrier 0 grace stop 0 bus error
        0 output buffer failures, 0 output buffers swapped out
```

第六步：如图 2-4 所示，设置 PC 机的 IP 地址并测试连通性。
如图 2-5 所示，使用 PING 测试连通性。

第七步：在 PC 机上 telnet 到路由器。
如图 2-6 所示，运行 telnet 192.168.2.1，出现如下结果：

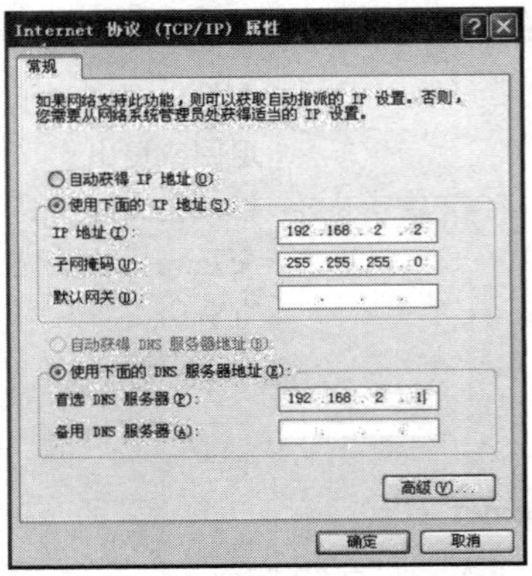

图 2-4　PC 机的 IP 地址配置

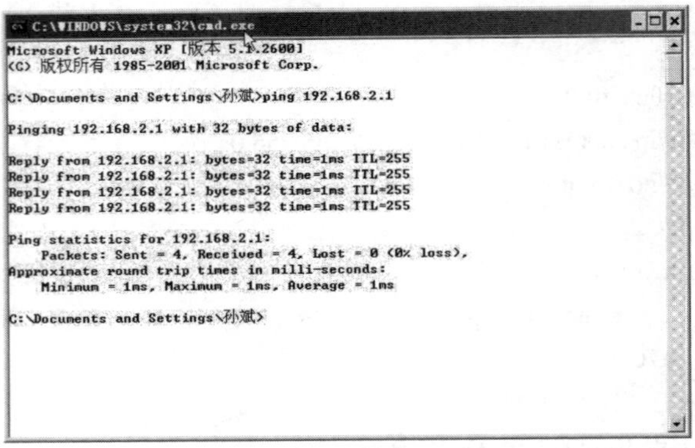

图 2-5　测试结果

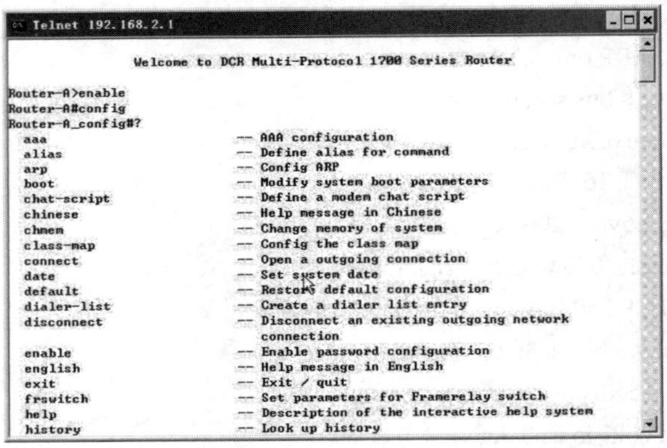

图 2-6　测试结果

带内远程的管理方法：Web 方式。

第八步：由于 1702 不支持 Web 管理方式，以下配置以 DCR-2611 为例，同前配置，将以太网接口地址配置为 192.168.2.2/24。

Router-C#config

Router-C_config#interface f0/0

Router-C_config_f0/0#ip address 192.168.2.2 255.255.255.0

Router-C_config_f0/0#no shutdown

```
Router-C_config_f0/0#exit
Router-C_config#ip http server                    ！开启 HTTP 服务
Router-C_config#^Z
Router-C#2089-2-22 01:22:06 Configured from console 0 by DEFAULT

Router-C#show running-config                      ！验证
Building configuration...

Current configuration:
!
!version 1.3.1S
service timestamps log date
service timestamps debug date
no service password-encryption
!
hostname Router-C
!
!
username abc password 0 digital
!
interface FastEthernet0/0
  ip address 192.168.2.2 255.255.255.0            ！F0/0 的 IP 地址
  no ip directed-broadcast
!
interface Async0/0
  no ip address
  no ip directed-broadcast
!
gateway-cfg
  Gateway keepAlive 60
  shutdown
!
ivr-cfg
!
ip http server                                    ！HTTP 服务开启状态
```

七、注意事项和排错

1. 在超级终端中的配置是对路由器的操作，这时的 PC 只是输入输出设备。
2. 在 telnet 和 Web 方式管理时，先测试连通性。

八、思考题

1. 带内和带外管理方式各有什么优点和缺点？

2. telnet 和 Web 的端口号是什么？

九、课后练习

请将所有设备的 IP 地址改为 10.0.0.0/24 这个网段的地址，将本实验重复配置。

实验三　路由器的基本配置

一、实验目的

1. 掌握路由器配置前的准备。
2. 掌握路由器的机器名的配置。
3. 接口 IP 地址、基本封装类型。

二、应用环境

1. 在执行配置之前，本节学习的基本配置是其他任务的基础。
2. 主要学习机器名、接口地址、特权模式密码等方法。

三、实验设备

1. DCR-1751　　两台。
2. CR-V35MT　　一条。
3. CR-V35FC　　一条。
4. 网线　　两条。

四、实验拓扑

该实验拓扑结构如图 2-7 所示。

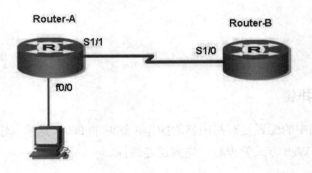

图 2-7　实验拓扑图

五、实验要求

该实验中路由器接口配置按表 2-3 所示。

表 2-3　路由器各接口配置表

Router-A			Router-B		
接口	类型	IP 地址	接口	类型	IP 地址
S1/1	DCE	192.168.2.1	S0/0	DTE	192.168.2.2
F0/0		192.168.2.1			

六、实验步骤

路由器 A 的基本配置：
第一步：恢复出厂设置。

Router>enable　　　　　　　　　　　　　　　　　　　　　　！进入特权模式
Router#2004-1-1 00:32:10 User DEFAULT enter privilege mode from console 0, level = 15
Router#show running-config　　　　　　　　　　　　　　　　！查看当前配置
Building configuration...
Current configuration:
!
!version 1.3.2E
<省略....>
Router#delete　　　　　　　　　　　　　　　　　　　　　　！删除配置文件
this file will be erased,are you sure?(y/n)y
Router#reboot　　　　　　　　　　　　　　　　　　　　　　！重新启动
Do you want to reboot the router(y/n)?y
Please wait…..

第二步：设置接口 IP 地址、DCE 的时钟频率以及验证。

Router>enable　　　　　　　　　　　　　　　　　　　　　　！进入特权模式
Router #config　　　　　　　　　　　　　　　　　　　　　　！进入全局配置模式
Router _config#hostname Router-A　　　　　　　　　　　　　！修改机器名
Router-A_config#interface s1/1　　　　　　　　　　　　　　！进入接口模式
Router-A_config_s1/0#ip address 192.168.2.1 255.255.255.0　！配置 IP 地址
Router-A_config_s1/0#physical-layer speed 64000　　　　　　！配置 DCE 时钟频率
Router-A_config_s1/0#no shutdown
Router-A_config_s1/0#^Z　　　　　　　　　　　　　　　　　！按 Ctrl＋Z 进入特权模式
Router-A#show interface s1/1　　　　　　　　　　　　　　　！查看接口状态
Serial1/0 is up, line protocol is down　　　　　　　　　　！对端没有配置，所以协议是 DOWN
　Mode=Sync DCE Speed=64000　　　　　　　　　　　　　　　！查看 DCE
　　DTR=UP,DSR=UP,RTS=UP,CTS=DOWN,DCD=UP
　Interface address is 192.168.2.1/24　　　　　　　　　　！查看 IP 地址
　MTU 1500 bytes, BW 64 kbit, DLY 2000 usec

　　　　Encapsulation prototol HDLC, link check interval is 10 sec　　　! 查看封装协议
Octets　Received0, Octets Sent 0
Frames Received 0, Frames Sent 0, Link-check Frames Received0
Link-check Frames Sent 89,　　LoopBack times 0
Frames Discarded 0, Unknown Protocols Frames Received 0, Sent failuile 0
　　Link-check Timeout 0, Queue Error 0, Link Error 0,
　　60 second input rate 0 bits/sec, 0 packets/sec!
　　60 second output rate 0 bits/sec, 0 packets/sec!
　　　0 packets input, 0 bytes, 8 unused_rx, 0 no buffer
　　　0 input errors, 0 CRC, 0 frame, 0 overrun, 0 ignored, 0 abort
　　　8 packets output, 192 bytes, 0 unused_tx, 0 underruns
　　error:
　　　0 clock, 0 grace
　　PowerQUICC SCC specific errors:
　　　　0 recv allocb mblk fail　　　0 recv no buffer
0 transmitter queue full　　　0 transmitter hwqueue_full

Router-A#config
Router-A_config#interface f0/0
Router-A_config_f0/0#ip address 192.168.2.1 255.255.255.0
Router-A_config_f0/0#no shutdown
Router-A_config_f0/0#^Z
Router-A#show interface f0/0
FastEthernet0/0 is up, line protocol is up
address is 00e0.0f18.1a70
　　Interface address is 192.168.2.1/24
　　MTU 1500 bytes, BW 100000 kbit, DLY 10 usec
　　Encapsulation ARPA, loopback not set
　　Keepalive not set
　　ARP type: ARPA, ARP timeout 04:00:00
　　60 second input rate 0 bits/sec, 0 packets/sec!
　　60 second output rate 6 bits/sec, 0 packets/sec!
　　Full-duplex, 100Mb/s, 100BaseTX, 1 Interrupt
　　　0 packets input, 0 bytes, 200 rx_freebuf
　　　Received 0 unicasts, 0 lowmark, 0 ri, 0 throttles
　　　0 input errors, 0 CRC, 0 framing, 0 overrun, 0 long
　　　1 packets output, 46 bytes, 50 tx_freebd, 0 underruns
　　　0 output errors, 0 collisions, 0 interface resets
　　　0 babbles, 0 late collisions, 0 deferred, 0 err600
　　　0 lost carrier, 0 no carrier 0 grace stop 0 bus error
　　　0 output buffer failures, 0 output buffers swapped out

第三步：设置特权模式密码。

Router-A_config#enable password 0 digitalchina　　　　　　　! 0 表示明文

```
Router-A_config#^Z
Router-A#2004-1-1 16:38:49 Configured from console 0 by DEFAULT
Router-A#exit
Router-A>enable                                         ! 再次进入特权模式
Password:                                               ! 需要输入密码
Access deny !
Router-A>enable
Password:                                               ! 注意输入时不显示
Router-A#2004-1-1 16:39:14 User DEFAULT enter privilege mode from console 0, level = 15
Router-A#
```

第四步：保存。

```
Router-A#write                                          ! 保存配置
Saving current configuration...
OK!
```

第五步：查看配置序列

```
Router-A#show running-config
Building configuration...

Current configuration:
!
!version 1.3.2E
service timestamps log date
service timestamps debug date
no service password-encryption
!
hostname Router-A                                       ! 查看机器名
!
enable password 0 digitalchina level 15                 ! 注意密码可以显示
!
interface FastEthernet0/0
  ip address 192.168.2.1 255.255.255.0                  ! 查看 IP 地址
  no ip directed-broadcast
!
<省略....>
interface Serial1/1
  ip address 192.168.2.1 255.255.255.0                  ! 查看 IP 地址
  no ip directed-broadcast
  physical-layer speed 64000
!
interface Async0/0
  no ip address
```

no ip directed-broadcast
　　!

路由器 B 的配置(命令解释参照路由器 A 的配置)：
第一步：恢复出厂设置。

Router>enable　　　　　　　　　　　　　　　　　　　　！进入特权模式
Router#2004-1-1 00:32:10 User DEFAULT enter privilege mode from console 0, level = 15

Router#show running-config　　　　　　　　　　　　　　！查看当前配置
Building configuration...

Current configuration:
!
!version 1.3.2E
<省略....>
Router#delete　　　　　　　　　　　　　　　　　　　　！删除配置文件
this file will be erased,are you sure?(y/n)y
Router#reboot　　　　　　　　　　　　　　　　　　　　！重新启动
Do you want to reboot the router(y/n)?y
Please wait…..

第二步：设置 IP 地址及验证。

Router>enable
Router#2004-1-1 01:04:14 User DEFAULT enter privilege mode from console 0, level = 15
Router#config
Router_config#hostname Router-B
Router-B_config#interface s1/0
Router-B_config_s1/0#ip address 192.168.2.2 255.255.255.0
Router-B_config_s1/0#no shutdown
Router-B_config_s1/0#^Z
Router-B#show interface s1/0
Serial1/0 is up, line protocol is up　　　　　　　　　　！此时接口和协议都是 up 状态
　Mode=Sync DTE
　　DTR=UP,DSR=UP,RTS=DOWN,CTS=UP,DCD=UP
　　Interface address is 192.168.2.2/24
　　MTU 1500 bytes, BW 64 kbit, DLY 2000 usec
　　Encapsulation prototol HDLC, link check interval is 10 sec
Octets　　Received0, Octets Sent 0
Frames Received 0, Frames Sent 0, Link-check Frames Received0
Link-check Frames Sent 391,　　LoopBack times 0
Frames Discarded 0, Unknown Protocols Frames Received 0, Sent failuile 0
　　Link-check Timeout 0, Queue Error 0, Link Error 0,
　　60 second input rate 0 bits/sec, 0 packets/sec!
　　60 second output rate 0 bits/sec, 0 packets/sec!

     0 packets input, 0 bytes, 8 unused_rx, 0 no buffer
     0 input errors, 0 CRC, 0 frame, 0 overrun, 0 ignored, 0 abort
     8 packets output, 192 bytes, 0 unused_tx, 0 underruns
  error:
     0 clock, 0 grace
  PowerQUICC SCC specific errors:
     0 recv allocb mblk fail   0 recv no buffer
0 transmitter queue full   0 transmitter hwqueue_full

第三步：保存。

Router-B#write
Saving current configuration...
OK!

第四步：查看配置序列。

Router-B#show running-config
Building configuration...

Current configuration:
!
!version 1.3.2E
service timestamps log date
service timestamps debug date
no service password-encryption
!
hostname Router-B
!
interface FastEthernet0/0
  ip address 192.168.3.1 255.255.255.0
  no ip directed-broadcast
!
interface Ethernet1/1
  no ip address
  no ip directed-broadcast
  duplex half
!
interface Serial1/0
  ip address 192.168.2.2 255.255.255.0
  no ip directed-broadcast
!
interface Async0/0
  no ip address
  no ip directed-broadcast
!

!
!
!
!
!
!

第五步：测试连通性。

Router-B#ping 192.168.2.1   ! PING Router-A 的地址
PING 192.168.2.1 (192.168.2.1): 56 data bytes
!!!!!
--- 192.168.2.1 ping statistics ---
5 packets transmitted, 5 packets received, 0% packet loss
round-trip min/avg/max = 20/22/30 ms

七、注意事项和排错

1. CR-V35FC 所连的接口为 DCE，需要配置时钟频率，CR-V35MT 所连的接口为 DTE。
2. 查看接口状态，如果接口是 DOWN，通常是线缆故障；如果协议是 DOWN，通常是时钟频率没有配，或者是两端封装协议不一致(封装的实验请参看实验四)。

八、配置序列

在步骤中已经列出。

九、思考题

1. 如果要将特权模式密码用密文显示，用什么参数？(请用？查看)
2. 如果在插槽 2 上的第 2 个快速以太网接口怎么表示？(f1/1)

十、课后练习

请将所有地址改为 10.0.0.0/24 这个网段，重复以上配置。

十一、相关命令详解

show interface
使用 show interface 全局配置命令配置接口状态。
show interface
show interface type interface-number
show interface type slot/port (用于带有非信道化 E1 的物理接口的路由器)
show interface serial slot/port:channel-group(用于显示非信道化 E1 的物理接口)
show interface serial slot/port.subinterface-number(用于显示子接口)
参数：实验参数如表 2-4 所示。

表 2-4 参数说明

参　　数	参　数　说　明
type	指定要配置的接口类型
interface-number	逻辑接口序号
slot	插槽或插卡编号
port	插槽或插卡端口编号
channel-group	范围为 0~30 的 E1 信道组号，使用 channel-group 配置命令定义
subinterface-number	范围为 1~32767 的子接口号

默认：无。

命令模式：管理态。

使用说明：若 show interface 命令后面不带任何参数，则显示所有接口的信息。

路由器 C 的基本配置(命令解释参照路由器 A 的配置)。

第一步：恢复出厂设置。

Router>enable　　　　　　　　　　　　　　　　　　！进入特权模式
Router#2004-1-1 00:32:10 User DEFAULT enter privilege mode from console 0, level = 15

Router#show running-config　　　　　　　　　　　　！查看当前配置
Building configuration...

Current configuration:
!
!version 1.3.2E
<省略....>

Router#delete　　　　　　　　　　　　　　　　　　！删除配置文件
this file will be erased,are you sure?(y/n)y
Router#reboot　　　　　　　　　　　　　　　　　　！重新启动
Do you want to reboot the router(y/n)?y
Please wait…..

第二步：设置 IP 地址及验证。

Router#config
Router_config#hostname Router-C
Router-C_config#interface f0/0
Router-C_config_f0/0#ip address 192.168.2.2 255.255.255.0
Router-C_config_f0/0#no shutdown
Router-C_config_f0/0#interface e1/0
Router-C_config_e1/0#ip address 192.168.3.2 255.255.255.0
Router-C_config_e1/0#no shutdown
Router-C_config_e1/0#^Z
Router-C#show interface

FastEthernet0/0 is up, line protocol is up
address is 00e0.0f20.0368
   Interface address is 192.168.2.2/24
   MTU 1500 bytes, BW 100000 kbit, DLY 10 usec
   Encapsulation ARPA, loopback not set
   Keepalive not set
   ARP type: ARPA, ARP timeout 04:00:00
   60 second input rate 0 bits/sec, 0 packets/sec!
   60 second output rate 0 bits/sec, 0 packets/sec!
   Full-duplex, 100Mb/s, 100BaseTX, 1 Interrupt
     0 packets input, 0 bytes, 200 rx_freebuf
     Received 0 unicasts, 0 lowmark, 0 ri, 0 throttles
     0 input errors, 0 CRC, 0 framing, 0 overrun, 0 long
     1 packets output, 46 bytes, 50 tx_freebd, 0 underruns
     0 output errors, 0 collisions, 0 interface resets
     0 babbles, 0 late collisions, 0 deferred, 0 err600
     0 lost carrier, 0 no carrier 0 grace stop 0 bus error
     0 output buffer failures, 0 output buffers swapped out
Ethernet1/0 is up, line protocol is up
address is 00e0.0f20.0369
   Interface address is 192.168.3.2/24
   MTU 1500 bytes, BW 10000 kbit, DLY 100 usec
   Encapsulation ARPA, loopback not set
   Keepalive not set
   ARP type: ARPA, ARP timeout 04:00:00
   60 second input rate 0 bits/sec, 0 packets/sec!
   60 second output rate 0 bits/sec, 0 packets/sec!
   Half-duplex, 10Mb/s, 10BaseTX, 0 Interrupt
     0 packets input, 0 bytes, 100 rx_freebuf
     Received 0 unicasts, 0 lowmark, 0 ri, 0 rx_busy
     0 input errors, 0 CRC, 0 framing, 0 overrun
     0 long, 0 i_collisions, 0 discard, 0 no buffer
     0 packets output, 0 bytes, 50 tx_freebd, 0 underruns
     0 output errors, 0 o_collisions, 0 late collisions
     0 lost carrier, 0 output buffer failures
Async0/0 is down, line protocol is down
   Hardware is Aux(PC16x50) Mode=Async Speed=9600
   DTR=UP,DSR=DOWN,RTS=UP,CTS=DOWN,DCD=DOWN
   MTU 1500 bytes, BW 9 kbit, DLY 10000 usec
   Encapsulation PPP, loopback not set
   Keepalive set(10 sec)
   LCP   Listening -- waiting for remote host to attempt open
   60 second input rate 0 bits/sec, 0 packets/sec!
   60 second output rate 0 bits/sec, 0 packets/sec!

pc16x50 UART 0, 5417 Interrupt
    0 packets input, 0 bytes, 0 no buffer
    0 input errors, 0 rx_dump, 0 Parity, 0 frame, 0 overrun
    0 packets output, 0 bytes, 0 underruns
    aux 0 output queue full, 0 frame has mblk more than one
flow control mode: hardware

第三步：保存。

Router-C#write
Saving current configuration...
OK!

第四步：查看配置序列。

Router-C#show running-config
Building configuration...

Current configuration:
!
!version 1.3.1S
service timestamps log date
service timestamps debug date
no service password-encryption
!
hostname Router-C
!
interface FastEthernet0/0
  ip address 192.168.2.2 255.255.255.0
  no ip directed-broadcast
!
interface Ethernet1/0
  ip address 192.168.3.2 255.255.255.0
  no ip directed-broadcast
  duplex half
!
interface Async0/0
  no ip address
  no ip directed-broadcast
!
!
gateway-cfg
  Gateway keepAlive 60
  shutdown
!
!

ivr-cfg
!
Router-C#

实验四 路由器的文件维护

一、实验目的

1. 掌握路由器的软件升级。
2. 备份和还原路由器的配置文件。

二、应用环境

1. 路由器生产厂家会不断推出新的软件版本，增加新的功能，管理员需要及时升级。
2. 配置文件需要及时备份，当文件损坏，或者设备更换时可以快速恢复。
3. 正常情况下可以通过TFTP或者FTP方式恢复，当设备无法正常启动时可以通过ZMODEN方式恢复。

三、实验设备

1. DCR-1751   一台。
2. PC 机   一台。
3. TFTP、FTP 软件。

四、实验拓扑

该实验拓扑结构如图 2-8 所示。

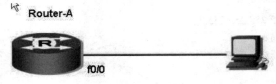

图 2-8 实验拓扑图

五、实验要求

该实验中各设备接口配置如表 2-5 所示。

表 2-5 配置表

Router-A		PC	
接口	IP 地址	网卡	IP 地址
F0/0	192.168.2.1		192.168.2.10

六、实验步骤

1. TFTP 方式(采用 UDP 协议，适合本地操作)

第一步：设置 PC 网卡地址为 192.168.2.10，并安装 3Cdaeom 软件，设置 TFTP SERVER 配置，如图 2-9 所示。

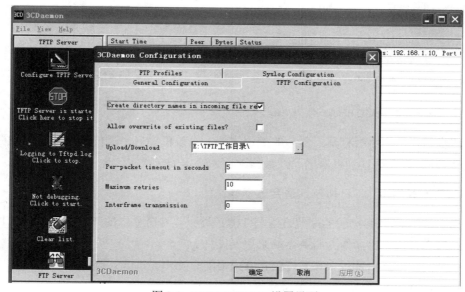

图 2-9　TFTP SERVER 设置界面

第二步：参照实验三，设置 DCR-1751 的 F0/0 接口地址为 192.168.2.1，并测试连通性。

Router-A#config
Router-A_config#interface f0/0
Router-A_config_f0/0#ip address 192.168.2.1 255.255.255.0
Router-A_config_f0/0#no shutdown
Router-A_config_f0/0#^Z
Router-A#show interface f0/0
FastEthernet0/0 is up, line protocol is up
address is 00e0.0f18.1a70
 Interface address is 192.168.2.1/24
 MTU 1500 bytes, BW 100000 kbit, DLY 10 usec
 Encapsulation ARPA, loopback not set
 Keepalive not set
<省略....>
Router-A#ping 192.168.2.10　　　　　　　　　　　　! PING PC 的地址
PING 192.168.2.10 (192.168.2.1): 56 data bytes
!!!!!
--- 192.168.2.1 ping statistics ---
5 packets transmitted, 5 packets received, 0% packet loss
round-trip min/avg/max = 20/22/30 ms

第三步：查看路由器文件，并将配置文件下载到 TFTP 服务器上。

Router-A#dir
Directory of /:
2   DCR-1751.bin   <FILE>   3589526   Sun Feb 7 06:28:15 2106
3   startup-config   <FILE>   516   Thu Jan 1 00:03:09 2004
free space 4751360

Router-A#copy flash:startup-config tftp: ！上传配置文件作为备份
Remote-server ip address[]?192.168.2.10   ！TFTP 服务器的 IP 地址
Destination file name[startup-config]?   ！默认的文件名
#
TFTP:successfully send 2 blocks ,516 bytes

第四步：使用写字板打开下载后的配置文件，修改机器名，上传到路由器中，重新启动后通过 show 命令观察到机器名已经被修改。

2. FTP 方式(采用 TCP 协议，适合远程操作)

路由器作为 FTP 客户端的配置方法与 TFTP 类似，将命令中的 TFTP 换成 FTP 即可；现在我们来学习将路由器作为服务器端来更新文件。

第一步：配置接口地址，参看 TFTP 方式中的第二步。

第二步：开启 FTP 服务，设置账号密码。

Router-A(Config)#ftp-server enable   ！开启 FTP 服务
Router-A(Config)#ip ftp username router   ！设置 FTP 账号
Router-A(Config)#ip ftp password 0 digitalchina   ！设置 FTP 密码

第三步：在 PC 机上打开 IE 作为客户端上传或下载。
在地址栏输入网址：ftp://router:digitalchina@192.168.2.1。

3. 启动到 monitor

当路由器的软件被破坏无法启动时，可以在启动过程中按 Ctrl + Break 组合键，启动到 monitor 模式中，使用 ZMODEM 方式恢复文件，所谓 ZMODEM 方式是从路由器的 Console 端口以波特率规定的速率通过 PC 的串口传输文件的一种方式，不需要网线。

第一步：将路由器重启，在启动过程中按 Ctrl + Break 组合键，启动到 MONITOR。

System Bootstrap, Version 0.1.8
Serial num:8IRT01V11B01000054 ,ID num:000847
Copyright (c) 1996-2000 by China Digitalchina CO.LTD
DCR-1700 Processor MPC860T @ 50Mhz
The current time: 2067-9-12 4:44:13
            Welcome to DCR Multi-Protool 1700 Series Router
monitor#

第二步：设置传输方式。

monitor#download c0 routerb　　　　　　　　　　　　！设置从 Console 传输文件
Speed: [9600]　　　　　　　　　　　　　　　　　　　！选择波特率

第三步：

(1) 打开 Windows 系统，选择"开始"｜"程序"｜"附件"｜"通讯"｜"超级终端"命令。

(2) 在"超级终端"窗口，打开"传送"菜单，选择"发送文件"命令，将打开"发送文件"对话框，如图 2-10 所示。

图 2-10　"发送文件"对话框

(3) 单击"浏览"按钮，接下来手动选择原备份的路由器软件，最后单击"发送"按钮，就会出现如图 2-11 所示的"正在发送"界面。

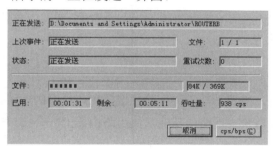

图 2-11　"正在发送"界面

4. 恢复遗失的密码

当密码遗忘，可以进入 monitor 模式，执行以下操作清除密码。

monitor#nopasswd

七、注意事项和排错

1. 路由器和 PC 直接相连时使用交叉线。
2. 关闭 PC 机上的防火墙。
3. 在实际工作中，通常使用日期或功能等标明配置文件。
4. 使用 ZMODEM 方式恢复文件，当文件比较大时比较耗时。

八、配置序列

无。

九、思考题

1. 请问 TFTP 和 FTP 这两种方式有什么区别？
2. 在 monitor 模式下，为什么不能使用 TFTP 方式？

十、课后练习

请写出使用 FTP 上传配置文件的过程。

实验五　单臂路由实验

一、实验目的

1. 了解路由器各接口的外观。
2. 掌握接口的功能。
3. 掌握接口的表示方法。

二、应用环境

路由器是三层设备，主要功能是进行路径选择和广域网的连接。而单臂路由实验主要是通过路由器的一个接口来实现路由功能，使用路由器上一个接口来实现二层交换机多个 VLAN 之间的互相通信。

三、实验设备

1. DCR 1702 或 2631　　一台。
2. DCS-3926S 交换机　　一台。
3. PC 机　　两台。
4. 直通网线若干。

四、实验拓扑

该实验拓扑结构如图 2-12 所示。

五、实验要求

如图 2-12 所示进行连线配置，PC1 到 PC2 可通。

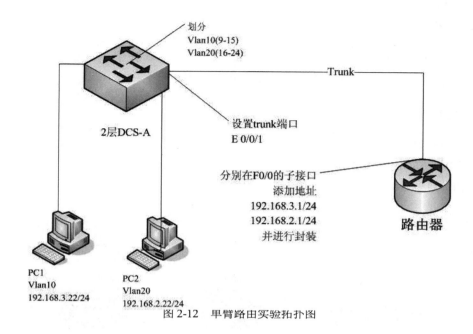

图 2-12 单臂路由实验拓扑图

六、重要配置

交换机配置：略。

路由器配置：

直接进入 F0/0.10 和 F0/0.20 子接口配置 IP 并进行 dot1q 封装。

Router#conf
Router_config#int f 0/0.10
Router_config_f0/0.10#ip ad 192.168.3.1 255.255.255.0
Router_config_f0/0.10#encapsulation dot1q 10
Router_config_f0/0.10#exit
Router_config#int f 0/0.20
Router_config_f0/0.20#ip ad 192.168.2.1 255.255.255.0
Router_config_f0/0.20#encapsulation dot1q 20
Router_config_f0/0.20#exit

实验：验证 PC1------ping-------PC2　　通。

实验六　路由器静态路由配置

一、实验目的

1. 理解路由表。
2. 掌握静态路由的配置。

二、应用环境

1. 在小规模环境里，静态路由是最佳的选择。
2. 静态路由开销小，但不灵活，适用于相对稳定的网络。

三、实验设备

1. DCR-1751   3 台。
2. CR-V35FC   一条。
3. CR-V35MT   一条。

四、实验拓扑

该实验拓扑结构如图 2-13 所示。

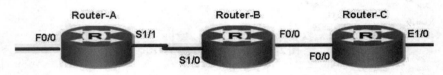

图 2-13 实验拓扑图

五、实验要求

该实验中各设备接口配置如表 2-6 所示。

表 2-6 配置表

Router-A		Router-B		Router-C	
S1/1(DCE)	192.168.5.1	S1/0(DTE)	192.168.5.2	F0/0	192.168.2.2
F0/0	192.168.0.1	F0/0	192.168.2.1	E1/0	192.168.3.1

六、实验步骤

第一步：参照实验三，按照表 2-6 配置所有接口的 IP 地址，保证所有接口全部是 up 状态，测试连通性。

第二步：查看 Router-A 的路由表。

```
Router-A#show ip route
Codes: C - connected, S - static, R - RIP, B - BGP, BC - BGP connected
       D - DEIGRP, DEX - external DEIGRP, O - OSPF, OIA - OSPF inter area
       ON1 - OSPF NSSA external type 1, ON2 - OSPF NSSA external type 2
       OE1 - OSPF external type 1, OE2 - OSPF external type 2
       DHCP - DHCP type

VRF ID: 0
```

C	192.168.0.0/24	is directly connected, FastEthernet0/0	！直连的路由
C	192.168.5.0/24	is directly connected, Serial1/1	！直连的路由

第三步：查看 Router-B 的路由表。

Router-B#show ip route
Codes: C - connected, S - static, R - RIP, B - BGP, BC - BGP connected
　　　　D - DEIGRP, DEX - external DEIGRP, O - OSPF, OIA - OSPF inter area
　　　　ON1 - OSPF NSSA external type 1, ON2 - OSPF NSSA external type 2
　　　　OE1 - OSPF external type 1, OE2 - OSPF external type 2
　　　　DHCP - DHCP type

VRF ID: 0

C	192.168.5.0/24	is directly connected, Serial1/0
C	192.168.2.0/24	is directly connected, FastEthernet0/0

第四步：查看 Router-C 的路由表。

Router-B#show ip route
Codes: C - connected, S - static, R - RIP, B - BGP, BC - BGP connected
　　　　D - DEIGRP, DEX - external DEIGRP, O - OSPF, OIA - OSPF inter area
　　　　ON1 - OSPF NSSA external type 1, ON2 - OSPF NSSA external type 2
　　　　OE1 - OSPF external type 1, OE2 - OSPF external type 2
　　　　DHCP - DHCP type

VRF ID: 0

C	192.168.5.0/24	is directly connected, Serial1/0
C	192.168.2.0/24	is directly connected, FastEthernet0/0

第五步：在 Router-A 上 ping 路由器 C。

Router-A#ping 192.168.2.2
PING 192.168.2.2 (192.168.2.2): 56 data bytes
.....
--- 192.168.2.2 ping statistics ---
5 packets transmitted, 0 packets received, 100% packet loss　　　！不通

第六步：在路由器 A 上配置静态路由。

Router-A#config
Router-A_config#ip route 192.168.2.0 255.255.255.0 192.168.5.2　　！配置目标网段和下一跳
Router-A_config#ip route 192.168.3.0 255.255.255.0 192.168.5.2

第七步：查看路由表。

```
Router-A#show ip route
Codes: C - connected, S - static, R - RIP, B - BGP, BC - BGP connected
       D - DEIGRP, DEX - external DEIGRP, O - OSPF, OIA - OSPF inter area
       ON1 - OSPF NSSA external type 1, ON2 - OSPF NSSA external type 2
       OE1 - OSPF external type 1, OE2 - OSPF external type 2
       DHCP - DHCP type

VRF ID: 0

C     192.168.0.0/24        is directly connected, FastEthernet0/0
C     192.168.5.0/24        is directly connected, Serial1/1
S     192.168.2.0/24        [1,0] via 192.168.5.2          ！注意静态路由的管理距离是 1
S     192.168.3.0/24        [1,0] via 192.168.5.2
```

第八步：配置路由器 B 的静态路由并查看路由表。

```
Router-B#config
Router-B_config#ip route 192.168.0.0 255.255.255.0 192.168.5.1
Router-B_config#ip route 192.168.3.0 255.255.255.0 192.168.2.2
Router-B_config#^Z
Router-B#show ip route
Codes: C - connected, S - static, R - RIP, B - BGP, BC - BGP connected
       D - DEIGRP, DEX - external DEIGRP, O - OSPF, OIA - OSPF inter area
       ON1 - OSPF NSSA external type 1, ON2 - OSPF NSSA external type 2
       OE1 - OSPF external type 1, OE2 - OSPF external type 2
       DHCP - DHCP type

VRF ID: 0

S     192.168.0.0/24        [1,0] via 192.168.5.1
C     192.168.5.0/24        is directly connected, Serial1/0
C     192.168.2.0/24        is directly connected, FastEthernet0/0
S     192.168.3.0/24        [1,0] via 192.168.2.2
```

第九步：配置路由器 C 的静态路由并查看路由表。

```
Router-C#config
Router-C_config#ip route 192.168.0.0 255.255.0.0 192.168.2.1      ！采用超网的方法
Router-C_config#^Z
Router-C#show ip route
Codes: C - connected, S - static, R - RIP, B - BGP
       D - DEIGRP, DEX - external DEIGRP, O - OSPF, OIA - OSPF inter area
       ON1 - OSPF NSSA external type 1, ON2 - OSPF NSSA external type 2
       OE1 - OSPF external type 1, OE2 - OSPF external type 2

S     192.168.0.0/16        [1,0] via 192.168.2.1          ！注意掩码是 16 位
```

C   192.168.2.0/24   is directly connected,   FastEthernet0/0
C   192.168.3.0/24   is directly connected,   Ethernet1/0

第十步：测试。

Router-C#ping 192.168.0.1
PING 192.168.0.1 (192.168.0.1): 56 data bytes
!!!!!   ！成功
--- 192.168.0.1 ping statistics ---
5 packets transmitted, 5 packets received, 0% packet loss
round-trip min/avg/max = 30/32/40 ms

七、注意事项和排错

1. 非直连的网段都要配置路由。
2. 以太网接口要接主机或交换机才能 up。
3. 串口注意 DCE 和 DTE 的问题。

八、配置序列

路由器 B 的序列

Router-B#show running-config
Building configuration...

Current configuration:
!
!version 1.3.2E
service timestamps log date
service timestamps debug date
no service password-encryption
!
hostname Router-B
!
interface FastEthernet0/0
  ip address 192.168.2.1 255.255.255.0
  no ip directed-broadcast
!
interface Serial1/0
  ip address 192.168.5.2 255.255.255.0
  no ip directed-broadcast
!
interface Async0/0
  no ip address
  no ip directed-broadcast
!

!
!
!
ip route 192.168.0.0 255.255.255.0 192.168.5.1
ip route 192.168.3.0 255.255.255.0 192.168.2.2
!
!

九、思考题

1. 什么情况下可以采用路由器 C 的超网配置方法？
2. 为什么只有当所有路由器都配置了路由以后才能通？
3. 静态路由有什么优势？在什么情况下使用？

实训八　多台路由器之间静态路由配置

一、实训设备

(1) DCS-3926S交换机　　2台。
(2) DCR-1702E 路由器　　1台；DCR-2631 路由器　　1台。
(3) PC机　　4台。
(4) 直通网线　　7根。

二、实训拓扑

该实验拓扑结构如图 2-14 所示。

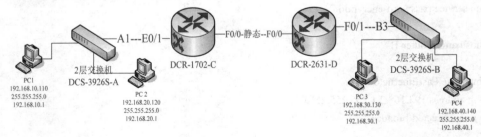

图 2-14　网络拓扑结构

三、实训要求及步骤

1. 该实验拓扑结构如图 2-13 所示。
2. 具体要求：按照图 2-13 所示进行连线，各个设备做如下配置。
(1) 交换机 A 划分 vl 10(9-12)，vl 20(13-16)，Trunk (1)；

(2) 交换机 B 划分 vl 30(17-20)，vl 40(21-24)，Trunk (3)。
(3) 设置路由器 C：

E 0/1.10	192.168.10.1/24	封装 vl 10;
E 0/1.20	192.168.20.1/24	封装 vl 20;
F 0/0	192.168.100.1/30	

(4) 设置路由器 D：

F 0/1.30	192.168.30.1/24	封装 vl 30;
F 0/1.40	192.168.40.1/24	封装 vl 40;
F 0/0	192.168.100.2/30	

(5) 配置 RC 的静态路由：

RC_config#ip route 192.168.30.0 255.255.255.0 192.168.100.2
RC_config#ip route 192.168.40.0 255.255.255.0 192.168.100.2

(6) 配置 RD 的静态路由：

RD_config#ip route 192.168.10.0 255.255.255.0 192.168.100.1
RD_config#ip route 192.168.20.0 255.255.255.0 192.168.100.1

(7) 根据连线位置正确配置 PC1、PC2 的地址(注意配上网关地址以及与 vlan 的对应关系)。

3. 实验结果：PC1---ping---PC2、3、4   通。
4. 查看设备状态：show vlan   show run   show ip route。

实验七 路由器 RIP-1 配置

一、实验目的

1. 掌握动态路由的配置方法。
2. 理解 RIP 协议的工作过程。

二、应用环境

1. 在路由器较多的环境里，手工配置静态路由给管理员带来较大的工作负担。
2. 在不太稳定的网络环境里，手工修改表不现实。

三、实验设备

1. DCR-1751　　三台。
2. CR-V35FC　　一条。
3. CR-V35MT　　一条。

四、实验拓扑

该实验拓扑结构如图 2-15 所示。

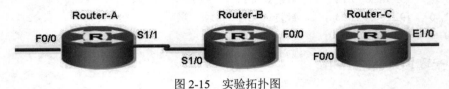

图 2-15　实验拓扑图

五、实验要求

该实验中各项配置如表 2-7 所示。

表 2-7　配置表

Router-A		Router-B		Router-C	
S1/1(DCE)	192.168.5.1	S1/0(DTE)	192.168.5.2	F0/0	192.168.2.2
F0/0	192.168.0.1	F0/0	192.168.2.1	E1/0	192.168.3.1

六、实验步骤

第一步：参照实验三，按照表 2-7 配置所有接口的 IP 地址，保证所有接口全部是 up 状态，测试连通性。

第二步：查看 Router-A 的路由表。

```
Router-A#show ip route
Codes: C - connected, S - static, R - RIP, B - BGP, BC - BGP connected
       D - DEIGRP, DEX - external DEIGRP, O - OSPF, OIA - OSPF inter area
       ON1 - OSPF NSSA external type 1, ON2 - OSPF NSSA external type 2
       OE1 - OSPF external type 1, OE2 - OSPF external type 2
       DHCP - DHCP type

VRF ID: 0

C      192.168.0.0/24       is directly connected, FastEthernet0/0    ！直连的路由
C      192.168.5.0/24       is directly connected, Serial1/1          ！直连的路由
```

第三步：查看 Router-B 的路由表。

Router-B#show ip route
Codes: C - connected, S - static, R - RIP, B - BGP, BC - BGP connected
       D - DEIGRP, DEX - external DEIGRP, O - OSPF, OIA - OSPF inter area
       ON1 - OSPF NSSA external type 1, ON2 - OSPF NSSA external type 2
       OE1 - OSPF external type 1, OE2 - OSPF external type 2
       DHCP - DHCP type

VRF ID: 0

C   192.168.5.0/24   is directly connected, Serial1/0
C   192.168.2.0/24   is directly connected, FastEthernet0/0

第四步：查看 Router-C 的路由表。

Router-B#show ip route
Codes: C - connected, S - static, R - RIP, B - BGP, BC - BGP connected
       D - DEIGRP, DEX - external DEIGRP, O - OSPF, OIA - OSPF inter area
       ON1 - OSPF NSSA external type 1, ON2 - OSPF NSSA external type 2
       OE1 - OSPF external type 1, OE2 - OSPF external type 2
       DHCP - DHCP type

VRF ID: 0

C   192.168.5.0/24   is directly connected, Serial1/0
C   192.168.2.0/24   is directly connected, FastEthernet0/0

第五步：在 Router-A 上 ping 路由器 C。

Router-A#ping 192.168.2.2
PING 192.168.2.2 (192.168.2.2): 56 data bytes
......
--- 192.168.2.2 ping statistics ---
5 packets transmitted, 0 packets received, 100% packet loss   ！不通

第六步：在路由器 A 上配置 RIP 协议并查看路由表。

Router-A_config#router rip   ！启动 RIP 协议
Router-A_config_rip#network 192.168.0.0   ！宣告网段
Router-A_config_rip#network 192.168.5.0
Router-A_config_rip#^Z
Router-A#sh ip route
Codes: C - connected, S - static, R - RIP, B - BGP, BC - BGP connected
       D - DEIGRP, DEX - external DEIGRP, O - OSPF, OIA - OSPF inter area
       ON1 - OSPF NSSA external type 1, ON2 - OSPF NSSA external type 2
       OE1 - OSPF external type 1, OE2 - OSPF external type 2
       DHCP - DHCP type

VRF ID: 0

C	192.168.0.0/24	is directly connected, FastEthernet0/0
C	192.168.5.0/24	is directly connected, Serial1/1

注意到并没有出现 RIP 学习到的路由。

第七步：在路由器 B 上配置 RIP 协议并查看路由表。

Router-B_config#router rip
Router-B_config_rip#network 192.168.5.0
Router-B_config_rip#network 192.168.2.0
Router-B_config_rip#^Z
Router-B#2004-1-1 00:15:58 Configured from console 0 by DEFAULT
Router-B#show ip route
Codes: C - connected, S - static, R - RIP, B - BGP, BC - BGP connected
       D - DEIGRP, DEX - external DEIGRP, O - OSPF, OIA - OSPF inter area
       ON1 - OSPF NSSA external type 1, ON2 - OSPF NSSA external type 2
       OE1 - OSPF external type 1, OE2 - OSPF external type 2
       DHCP - DHCP type

VRF ID: 0

R	192.168.0.0/16	[120,1] via 192.168.5.1(on Serial1/0)	！从 A 学习到的路由
C	192.168.5.0/24	is directly connected, Serial1/0	
C	192.168.2.0/24	is directly connected, FastEthernet0/0	

第八步：在路由器 C 上配置 RIP 协议并查看路由表。

Router-C_config#router rip
Router-C_config_rip#network 192.168.2.0
Router-C_config_rip#network 192.168.3.0
Router-C_config_rip#^Z
Router-C#show ip route
Codes: C - connected, S - static, R - RIP, B - BGP
       D - DEIGRP, DEX - external DEIGRP, O - OSPF, OIA - OSPF inter area
       ON1 - OSPF NSSA external type 1, ON2 - OSPF NSSA external type 2
       OE1 - OSPF external type 1, OE2 - OSPF external type 2

R	192.168.0.0/16	[120,2] via 192.168.2.1(on	FastEthernet0/0)
R	192.168.5.0/24	[120,1] via 192.168.2.1(on	FastEthernet0/0)
C	192.168.2.0/24	is directly connected,	FastEthernet0/0
C	192.168.3.0/24	is directly connected,	Ethernet1/0

第九步：再次查看 A 和 B 的路由表。

Router-B#show ip route
Codes: C - connected, S - static, R - RIP, B - BGP, BC - BGP connected
       D - DEIGRP, DEX - external DEIGRP, O - OSPF, OIA - OSPF inter area
       ON1 - OSPF NSSA external type 1, ON2 - OSPF NSSA external type 2
       OE1 - OSPF external type 1, OE2 - OSPF external type 2
       DHCP - DHCP type

VRF ID: 0

R   192.168.0.0/16   [120,1] via 192.168.5.1(on Serial1/0)
C   192.168.5.0/24   is directly connected, Serial1/0
C   192.168.2.0/24   is directly connected, FastEthernet0/0
R   192.168.3.0/24   [120,1] via 192.168.2.2(on FastEthernet0/0)

Router-A#show ip route
Codes: C - connected, S - static, R - RIP, B - BGP, BC - BGP connected
       D - DEIGRP, DEX - external DEIGRP, O - OSPF, OIA - OSPF inter area
       ON1 - OSPF NSSA external type 1, ON2 - OSPF NSSA external type 2
       OE1 - OSPF external type 1, OE2 - OSPF external type 2
       DHCP - DHCP type

VRF ID: 0

C   192.168.0.0/24   is directly connected, FastEthernet0/0
C   192.168.5.0/24   is directly connected, Serial1/1
R   192.168.2.0/24   [120,1] via 192.168.5.2(on Serial1/1)
R   192.168.3.0/24   [120,2] via 192.168.5.2(on Serial1/1)
！注意到所有网段都学习到了路由

第十步：相关的查看命令。

Router-A#show ip rip   ！显示 RIP 状态
RIP protocol: Enabled
  Global version: default(Decided on the interface version control)
  Update: 30,   Expire: 180,   Holddown: 120
  Input-queue: 50
  Validate-update-source enable
  No neighbor

Router-A#sh ip rip protocol   ！显示协议细节
RIP is Active
  Sending updates every 30 seconds, next due in 30 seconds   ！注意定时器的值
  Invalid after 180 seconds, holddown 120

update filter list for all interfaces is:
update offset list for all interfaces is:
Redistributing:
Default version control: send version 1, receive version 1 2
　　Interface　　　　　Send　　　　　　　Recv
　　FastEthernet0/0　　1　　　　　　　　1 2
　　Serial1/1　　　　　1　　　　　　　　1 2
Automatic network summarization is in effect
Routing for Networks:
　　192.168.5.0/24
　　192.168.0.0/16
Distance: 120 (default is 120)　　　　　　　　　　! 注意默认的管理距离
　　Maximum route count: 1024,　　Route count:6

Router-A#show ip rip database　　　　　　　　　! 显示 RIP 数据库
　192.168.0.0/24　　directly connected　FastEthernet0/0
　192.168.0.0/24　　auto-summary
　192.168.5.0/24　　directly connected　Serial1/1
　192.168.5.0/24　　auto-summary
　192.168.2.0/24　　[120,1]　via 192.168.5.2 (on Serial1/1)　00:00:13　! 收到 RIP 广播的时间
　192.168.3.0/24　　[120,2]　via 192.168.5.2 (on Serial1/1)　00:00:13

Router-A#sh ip route rip　　　　　　　　　　　　! 仅显示 RIP 学习到的路由
　R　　192.168.2.0/24　　[120,1] via 192.168.5.2(on Serial1/1)
　R　　192.168.3.0/24　　[120,2] via 192.168.5.2(on Serial1/1)

七、注意事项和排错

1. 只能宣告直连的网段。
2. 宣告时不附加掩码。
3. 分配地址时最好是连续的子网，以免 RIP 汇聚出现错误。

八、配置序列

路由器 B 的序列

Router-B#show running-config
Building configuration...

Current configuration:
!
!version 1.3.2E
service timestamps log date
service timestamps debug date
no service password-encryption

```
!
hostname Router-B
!
interface FastEthernet0/0
  ip address 192.168.2.1 255.255.255.0
  no ip directed-broadcast
!
interface Serial1/0
  ip address 192.168.5.2 255.255.255.0
  no ip directed-broadcast
!
interface Async0/0
  no ip address
  no ip directed-broadcast
!
!
router rip
  network 192.168.2.0
  network 192.168.5.0
!
!
!
```

九、思考题

1. 为什么 B 没有配置 RIP 协议时，A 没有出现 RIP 路由？
2. 如果不是连续的子网，会出现什么结果？
3. RIP 的广播周期是多少？

实训九　多台路由器的动态 RIP-1 路由配置

一、实训设备

(1) DCS-3926S 交换机　　2 台。

(2) DCR-1702E 路由器　　1 台；DCR-2631 路由器　　1 台。

(3) PC 机　　4 台。

(4) 直通网线　　7 根。

二、实训拓扑

该实验拓扑结构如图 2-16 所示。

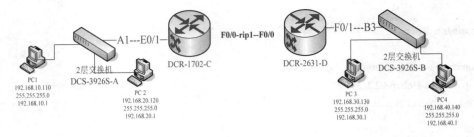

图 2-16　网络拓扑结构

三、实训要求及步骤

具体要求：按照图 2-16 所示进行连线，各个设备做如下配置。

(1) 交换机 A 划分 vl 10(9-12)，vl 20(13-16)，Trunk (1)；
(2) 交换机 B 划分 vl 30(17-20)，vl 40(21-24)，Trunk (3)。
(3) 设置路由器 C：

E 0/1.10　　192.168.10.1/24　　　　封装 vl 10；
E 0/1.20　　192.168.20.1/24　　　　封装 vl 20；
F 0/0　　　192.168.100.1/30

(4) 设置路由器 D：

F 0/1.30　　192.168.30.1/24　　　　封装 vl 30；
F 0/1.40　　192.168.40.1/24　　　　封装 vl 40；
F 0/0　　　192.168.100.2/30

(5) 启用路由器的动态 RIP 协议，配置 C 的动态路由 rip：

RC_config#router rip
RC_config_rip#network 192.168.10.0
RC_config_rip#network 192.168.20.0
RC_config_rip#network 192.168.100.0

(6) 配置 D 的动态路由 rip：

RD_config#router rip
RD_config_rip#network 192.168.30.0
RD_config_rip#network 192.168.40.0
RD_config_rip#network 192.168.100.0

验证：sh ip rip　　　　sh ip route。
(7) 根据连线位置正确配置 PC1、PC2 的地址(注意配上网关地址以及与 vlan 的对应关系)。

实验结果：PC1---ping---PC2、3、4　　　通。
查看设备状态：show vlan　　show run　　show ip route。

实验八　路由器 RIP-2 配置

一、实验目的

1. 掌握动态路由的配置方法。
2. 理解 RIP-2 协议的工作过程。
3. 理解 RIP-2 对变长子网掩码的支持。

二、应用环境

1. 在不连续的变长子网掩码环境中，RIP-1 的自动汇总会得出错误的路由。
2. RIP-2 可以关闭自动汇总，从而得出正确的路由。
3. 尽量不要分配不连续的子网。
4. 在路由器较多的环境里，手工配置静态路由给管理员带来大的工作负担。
5. 在不太稳定的网络环境里，手工修改表不现实。

三、实验设备

1. DCR-1751　　　3 台。
2. CR-V35FC　　　一条。
3. CR-V35MT　　　一条。

四、实验拓扑

该实验拓扑结构如图 2-17 所示。

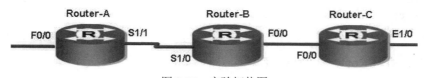

图 2-17　实验拓扑图

五、实验要求

该实验中各设备接口配置如表 2-8 所示。

表 2-8　配置表

Router-A		Router-B		Router-C	
S1/1(DCE)	192.168.5.1/24	S1/0(DTE)	192.168.5.2/24	F0/0	192.168.2.2/24
F0/0	172.16.3.1/24	F0/0	192.168.2.1/24	E1/0	172.16.4.1/24

六、实验步骤

第一步：参照实验三，按照表 2-8 配置所有接口的 IP 地址，保证所有接口全部是 up 状态，测试连通性。

第二步：配置 Router-A 的 RIP-1 协议。

Router-A_config#router rip
Router-A_config_rip# network 192.168.5.0
Router-A_config_rip#network   172.16.3.0 255.255.255.0   !在 RIP-1 里掩码是没有意义的
Router-A_config_rip#^Z

第三步：查看 Router-A 的路由表。

Router-A#show ip route
Codes: C - connected, S - static, R - RIP, B - BGP, BC - BGP connected
         D - DEIGRP, DEX - external DEIGRP, O - OSPF, OIA - OSPF inter area
         ON1 - OSPF NSSA external type 1, ON2 - OSPF NSSA external type 2
         OE1 - OSPF external type 1, OE2 - OSPF external type 2
         DHCP - DHCP type

VRF ID: 0

C   172.16.3.0/24   is directly connected, FastEthernet0/0   ! 直连的路由
C   192.168.5.0/24   is directly connected, Serial1/1   ! 直连的路由

第四步：配置 Router-B 的 RIP-1 协议。

Router-B_config#router rip
Router-B_config_rip#network 192.168.5.0
Router-B_config_rip#network 192.168.2.0
Router-B_config_rip#^Z

第五步：查看 Router-B 的路由表。

Router-B#show ip route
Codes: C - connected, S - static, R - RIP, B - BGP, BC - BGP connected
         D - DEIGRP, DEX - external DEIGRP, O - OSPF, OIA - OSPF inter area
         ON1 - OSPF NSSA external type 1, ON2 - OSPF NSSA external type 2
         OE1 - OSPF external type 1, OE2 - OSPF external type 2
         DHCP - DHCP type

VRF ID: 0

R   172.16.0.0/16   [120,1] via 192.168.5.1(on Serial1/0)   ! 注意是有类的地址
C   192.168.5.0/24   is directly connected, Serial1/0
C   192.168.2.0/24   is directly connected, FastEthernet0/0

第六步：配置路由器 C 的 RIP-1 协议。

Router-C_config#router rip
Router-C_config_rip#net 192.168.2.0
Router-C_config_rip#net 172.16.4.0 255.255.255.0   ！ RIP-1 是有类的路由协议
Router-C_config_rip#^Z

第七步：再次查看路由器 B 的路由表。

Router-B#show ip route
Codes: C - connected, S - static, R - RIP, B - BGP, BC - BGP connected
      D - DEIGRP, DEX - external DEIGRP, O - OSPF, OIA - OSPF inter area
      ON1 - OSPF NSSA external type 1, ON2 - OSPF NSSA external type 2
      OE1 - OSPF external type 1, OE2 - OSPF external type 2
      DHCP - DHCP type

VRF ID: 0

R   172.16.0.0/16   [120,1] via 192.168.5.1(on Serial1/0)
                            [120,1] via 192.168.2.2(on FastEthernet0/0)
                            ！ 由于有类路由的自动汇总，出现了错误的路由
C   192.168.5.0/24   is directly connected, Serial1/0
C   192.168.2.0/24   is directly connected, FastEthernet0/0

第八步：在所有路由器上配置 RIP-2 协议并关闭自动汇总。

Router-C_config#router rip
Router-C_config#version 2   ！ 指明为版本 2
Router-C_config# no auto-summary   ！ 关闭自动汇总

Router-B_config#router rip
Router-B_config_rip#version 2
Router-B_config_rip#no auto-summary

Router-A_config#router rip
Router-A_config_rip#version 2
Router-A_config_rip#no auto-summary

第九步：再次查看所有的路由表。
B 的路由表：

Router-B#sh ip route
Codes: C - connected, S - static, R - RIP, B - BGP, BC - BGP connected
      D - DEIGRP, DEX - external DEIGRP, O - OSPF, OIA - OSPF inter area
      ON1 - OSPF NSSA external type 1, ON2 - OSPF NSSA external type 2
      OE1 - OSPF external type 1, OE2 - OSPF external type 2

DHCP - DHCP type

VRF ID: 0

R	172.16.0.0/16	[120,16] via 192.168.2.2(on FastEthernet0/0)	
R	172.16.3.0/24	[120,1] via 192.168.5.1(on Serial1/0)	！正确的路由
R	172.16.4.0/24	[120,1] via 192.168.2.2(on FastEthernet0/0)	
C	192.168.5.0/24	is directly connected, Serial1/0	
C	192.168.2.0/24	is directly connected, FastEthernet0/0	

A 的路由表：

Router-A#sh ip route
Codes: C - connected, S - static, R - RIP, B - BGP, BC - BGP connected
       D - DEIGRP, DEX - external DEIGRP, O - OSPF, OIA - OSPF inter area
       ON1 - OSPF NSSA external type 1, ON2 - OSPF NSSA external type 2
       OE1 - OSPF external type 1, OE2 - OSPF external type 2
       DHCP - DHCP type

VRF ID: 0

C	172.16.3.0/24	is directly connected, FastEthernet0/0	
R	172.16.4.0/24	[120,2] via 192.168.5.2(on Serial1/1)	！正确的路由
C	192.168.5.0/24	is directly connected, Serial1/1	
R	192.168.2.0/24	[120,1] via 192.168.5.2(on Serial1/1)	

Router-C#sh ip route
Codes: C - connected, S - static, R - RIP, B - BGP
       D - DEIGRP, DEX - external DEIGRP, O - OSPF, OIA - OSPF inter area
       ON1 - OSPF NSSA external type 1, ON2 - OSPF NSSA external type 2
       OE1 - OSPF external type 1, OE2 - OSPF external type 2

R	172.16.3.0/24	[120,2] via 192.168.2.1(on FastEthernet0/0)
C	172.16.4.0/24	is directly connected, Ethernet1/0
R	192.168.5.0/24	[120,1] via 192.168.2.1(on FastEthernet0/0)
C	192.168.2.0/24	is directly connected, FastEthernet0/0

！注意到所有网段都学习到了正确掩码的路由

第十步：相关的查看命令。

Router-A#show ip rip   ！显示 RIP 状态
….
Router-A#show ip rip protocol   ！显示协议细节
….
Router-A#show ip rip database   ！显示 RIP 数据库

….
Router-A#show ip route rip   ！仅显示 RIP 学习到的路由
….

七、注意事项和排错

1. 只能宣告直连的网段。
2. 宣告时不附加掩码。
3. 分配地址时最好是连续的子网，以免 RIP 汇聚出现错误。

八、配置序列

路由器 B 的序列

Router-B#show running-config
Building configuration...

Current configuration:
!
!version 1.3.2E
service timestamps log date
service timestamps debug date
no service password-encryption
!
hostname Router-B
!
interface FastEthernet0/0
 ip address 192.168.2.1 255.255.255.0
 no ip directed-broadcast
!
interface Serial1/0
 ip address 192.168.5.2 255.255.255.0
 no ip directed-broadcast
!
interface Async0/0
 no ip address
 no ip directed-broadcast
!
!
router rip
 no auto-summary
 version 2
 network 192.168.2.0
 network 192.168.5.0
!

九、思考题

1. RIP-1 与 RIP-2 有什么不同？
2. 如果不是连续的子网，会出现什么结果？
3. RIP-2 的组播地址是什么？(可以通过 debug ip rip protocol 查看，注意及时使用 no debug all 关闭)。

十、相关命令详解

network

使用 network 命令为 RIP 协议指定连接的网络号，no network 则取消一个网络号。

network network-numbe <network-mask>

no network network-number <network-mask>

参数：实验参数如表 2-9 所示。

表 2-9 参数说明

参数	参数说明
Network-number	直接相连网络的网络 IP 地址
Network-mask	(可选)直接相连网络的网络 IP 地址掩码

默认：无网络被指定。

命令模式：路由配置态。

使用说明：指定的网络号不能包含任何子网信息。可以指定多个network命令。RIP更新只能在这个网络上的接口上发送和接收。

RIP 对指定网络上的接口发送 RIP 更新。如果一个接口相连的网络没有被指定，它也不会在任何 RIP 更新中被宣告。

举例：下面的例子定义了 RIP 作为与网络128.99.0.0和192.31.7.0相连接口的路由协议。

router rip
 network 128.99.0.0
 network 192.31.7.0

实训十 多台路由器的动态 RIP-2 路由配置

一、实训设备

(1) DCS-3926S 交换机   2 台。
(2) DCR-1702E 路由器   1 台；DCR-2631 路由器   1 台。
(3) PC 机   4 台。

(4) 直通网线　　7 根。

二、实训拓扑

该实验拓扑结构如图 2-18 所示。

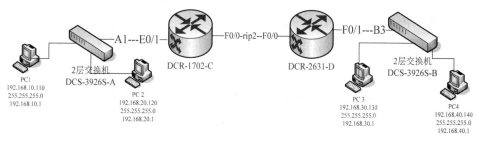

图 2-18　网络拓扑结构

三、实训要求及步骤

按照图 2-18 所示进行连线，各个设备做如下配置。

(1) 交换机 A 划分 vl 10(9-12)，vl 20(13-16)，Trunk (1)；

(2) 交换机 B 划分 vl 30(17-20)，vl 40(21-24)，Trunk (3)。

(3) 设置路由器 C：

E 0/1.10　　192.168.10.1/24　　封装 vl 10；
E 0/1.20　　192.168.20.1/24　　封装 vl 20；
F 0/0　　　192.168.100.1/30

(4) 设置路由器 D：

F 0/1.30　　192.168.30.1/24　　封装 vl 30；
F 0/1.40　　192.168.40.1/24　　封装 vl 40；
F 0/0　　　192.168.100.2/30

(5) 启用路由器的动态 RIP 协议，配置 C 的动态路由 rip：

RC_config#router rip

RC_config_rip#version 2

RC_config_rip#network 192.168.10.0

RC_config_rip#network 192.168.20.0

RC_config_rip#network 192.168.100.0

(6) 配置 D 的动态路由 rip：

RD_config#router rip

RD_config_rip#version 2

RD_config_rip#network 192.168.30.0

RD_config_rip#network 192.168.40.0

RD_config_rip#network 192.168.100.0

验证：sh ip rip   sh ip route。

(7) 根据连线位置正确配置 PC1、PC2 的地址(注意配上网关地址以及与 vlan 的对应关系)。

实验结果：PC1---ping---PC2、3、4   通。

查看设备状态：show vlan   show run   show ip route。

实验九 静态路由和直连路由引入配置

一、实验目的

1. 掌握路由引入的配置。
2. 理解路由引入的原理。

二、应用环境

在某些应用环境中需要将静态路由引入到某些动态路由协议里。

三、实验设备

DCR-1751   两台。

四、实验拓扑

该实验拓扑结构如图 2-19 所示。

图 2-19 实验拓扑图

五、实验要求

该实验中各设备接口配置如表 2-10 所示。

表 2-10 配置表

Router-A		Router-B	
S1/1(DCE)	192.168.5.1/24	S1/0	192.168.5.2/24
F0/0	192.16.0.1/24	F0/0	192.168.2.1/24

六、实验步骤

1. 引入到 RIP 协议中

第一步：参照实验三和上表配置接口地址并测试连通性。
第二步：配置路由器 A 的静态路由，查看直连和静态路由。

Router-A#config
Router-A_config#ip route 191.13.2.0 255.255.255.0 192.168.0.4　　　! 配置静态路由
Router-A_config#^Z
Router-A#sh ip route
Codes: C - connected, S - static, R - RIP, B - BGP, BC - BGP connected
　　　　D - DEIGRP, DEX - external DEIGRP, O - OSPF, OIA - OSPF inter area
　　　　ON1 - OSPF NSSA external type 1, ON2 - OSPF NSSA external type 2
　　　　OE1 - OSPF external type 1, OE2 - OSPF external type 2
　　　　DHCP - DHCP type

VRF ID: 0

S　　　191.13.2.0/24　　　[1,0] via 192.168.0.4
C　　　192.168.0.0/24　　　is directly connected, FastEthernet0/0
C　　　192.168.5.0/24　　　is directly connected, Serial1/1

第三步：在 A 上配置 RIP 协议，并将直连和静态路由引入。

Router-A_config#router rip
Router-A_config_rip#network 192.168.5.0　　　! 注意并没有宣告 192.168.0.0
Router-A_config_rip#redistribute connect　　　! 将直连的路由引入
Router-A_config_rip#redistribute static　　　! 将静态路由引入

第四步：在 B 上配置 RIP 协议，查看从 A 学习到的被引入的路由。

Router-B#conf
Router-B_config#router rip
Router-B_config_rip#network 192.168.5.0
Router-B_config_rip#^Z
Router-B#sh ip route
Codes: C - connected, S - static, R - RIP, B - BGP, BC - BGP connected
　　　　D - DEIGRP, DEX - external DEIGRP, O - OSPF, OIA - OSPF inter area
　　　　ON1 - OSPF NSSA external type 1, ON2 - OSPF NSSA external type 2
　　　　OE1 - OSPF external type 1, OE2 - OSPF external type 2
　　　　DHCP - DHCP type

VRF ID: 0

R　　　191.13.0.0/16　　　[120,1] via 192.168.5.1(on Serial1/0)　　! 注意是有类的路由

R	192.168.0.0/16	[120,1] via 192.168.5.1(on Serial1/0)
C	192.168.5.0/24	is directly connected, Serial1/0
C	192.168.2.0/24	is directly connected, FastEthernet0/0

2. 引入到 OSPF 协议中

第一步和第二步同上。

第三步：在 A 上配置 OSPF 协议，并将直连和静态引入。

```
Router-A#conf
Router-A_config#router ospf 1
Router-A_config_ospf_1#net 192.168.5.0 255.255.255.0 area 0
Router-A_config_ospf_1#redistribute connect
Router-A_config_ospf_1#redistribute static
```

第四步：在 B 上配置 OSPF 协议，并查看从 A 学习到的路由。

```
Router-B#conf
Router-B_config#router ospf 1
Router-B_config_ospf_1#net 192.168.5.0 255.255.255.0 area 0
Router-B_config_ospf_1#exit
Router-B_config#no router rip
Router-B_config#^Z
Router-B#sh ip route
Codes: C - connected, S - static, R - RIP, B - BGP, BC - BGP connected
       D - DEIGRP, DEX - external DEIGRP, O - OSPF, OIA - OSPF inter area
       ON1 - OSPF NSSA external type 1, ON2 - OSPF NSSA external type 2
       OE1 - OSPF external type 1, OE2 - OSPF external type 2
       DHCP - DHCP type

VRF ID: 0
```

O E2	191.13.2.0/24	[150,100] via 192.168.5.1(on Serial1/0)
！注意管理距离和花费值		
O E2	192.168.0.0/24	[150,100] via 192.168.5.1(on Serial1/0)
C	192.168.5.0/24	is directly connected, Serial1/0
C	192.168.2.0/24	is directly connected, FastEthernet0/0

七、注意事项和排错

1. 注意是将已经存在的路由引入到动态路由协议中，静态路由要先配置。
2. 引入成功后，该动态路由协议将引入的路由发布出去，在其他的路由器上查看。
3. RIP-1 是有类的路由协议。

八、配置序列

Router-A#sh run
Building configuration...

Current configuration:
!
!version 1.3.2E
service timestamps log date
service timestamps debug date
no service password-encryption
!
hostname Router-A
!
!
interface FastEthernet0/0
  ip address 192.168.0.1 255.255.255.0
  no ip directed-broadcast
!
interface Serial1/0
  no ip address
  no ip directed-broadcast
  physical-layer speed 64000
!
interface Serial1/1
  ip address 192.168.5.1 255.255.255.0
  no ip directed-broadcast
  physical-layer speed 64000
!
interface Async0/0
  no ip address
  no ip directed-broadcast
!
!
router ospf 1
  network 192.168.5.0 255.255.255.0 area 0
  redistribute static
  redistribute connect
!
!
ip route 191.13.2.0 255.255.255.0 192.168.0.4
!
!

九、思考题

1. 在配置 RIP 和 OSPF 协议时，为什么不宣告 192.168.0.0？
2. 如果配置静态路由时的下一跳是 192.168.5.2 会导致什么结果？

十、课后练习

请将静态路由：10.10.10.0/24、via 192.168.0.9 引入到 RIP 和 OSPF 中。

十一、相关命令详解

redistribute

使用 redistribute 路由器配置命令把路由从一个路由域重新分布到另一个路由域。使用 no redistribute 取消重新分布。

redistribute protocol [process-id] [route-map map-name]

no redistribute protocol [process-id] [route-map map-name]

参数：实验参数如表 2-11 所示。

表 2-11 参数说明

参数	参数说明
protocol	路由要被重新分布的源协议，它可以是下面几个关键词之一：bgp、ospf、static[ip]、connected 和 rip 关键词 static[ip]被用来重新分布 IP 静态路由。当路由重新分布到 IS-IS 中时，使用这个可选的 IP 关键词 关键词 connected 是指那些在接口上 IP 激活后自动建立起来的路由。对于像 OSPF 和 IS-IS 的路由协议，这些路由是被作为自治系统的外部路由被重新分布的
process-id	(可选项)对于 bgp 或 bigp，该参数是指 16 位数字的自治系统号 对于 OSPF，这是路由钥被重新分布的相应的 OSPF 进程 ID。这就标识了路由进程。它是一个非 0 的十进制数 对于 rip，并不需要进程标识 process-id
route-map	(可选项)该参数告诉路由映射对那些从源协议导入到当前路由协议的路由进行过滤。如果这个参数没有给出，所有路由将重新分布。如果给出这个关键词但没有列出路由映射标记，将没有路由被导入

默认：路由重新分布处于无效状态。

Protocol——无路由协议被定义。

process-id——无进程 ID 被定义。

route-map map-tag——如果参数 route-map 没有给出，所有路由将重新分布。如果没有输入 map-tag，没有路由被导入。

命令模式：路由配置态。

使用说明：改变或者使任何关键词无效将不会影响其他关键词的状态。

当路由器接收到一个带有内部路由权值的链路状态协议分组时，它会把从自己到那个重新分布路由器的权值与所宣告的到达目的的权值之和作为路由的权值，而外部的路由权值仅仅考虑在宣告中声明的到达目的的权值。

重新分布的路由信息永远会被 filter out 路由器配置命令所过滤，这可以保证只有管理员指定的路由可以进入接受的路由协议中。

不管什么时候使用 redistribute 或 default-information 路由器配置命令把路由重新分布到 OSPF 路由域内，那台路由器自动成为自治系统边界路由器 ASBR。但是，在默认情况下，ASBR 并不产生一条到 OSPF 路由域内的默认路由(default route)。

当路由在 OSPF 进程之间重新分布时，OSPF 路由权值都将使用。

当路由被重新分布到 OSPF 中，如果没有用 meric 关键词指定路由权值，对于来自其他所有协议(不包括 BGP)的路由，OSPF 使用 20 作为默认的路由权值(对于 BGP，使用路由权值 1)。进一步说，当路由在同一台路由器上两个 OSPF 进程之间重新分布时，如果没有指定默认的路由权值，那么一个路由进程内的路由权值会被带入到执行重新分布的进程中。

当路由重新分布到 OSPF 中时，如果关键词 subnets 没有给出，那么只有那些没有携带子网新的路由可以被重新分布。

被 redistribute 命令影响的 connected 路由是那些没有用 network 命令指定的路由。不能使用 default-metric 命令来影响宣告 connected 路由的权值。

注意：

在 redistribute 指定的路由权值一直用 default-metric 指定的路由权值。

除非 default-information originate 命令给出，否则不允许对从 IGP 或 EGP 到 BGP 的路由进行重新分布。

举例：下面的例子使得 OSPF 路由可以重新分布到 BGP 路由域中。

router bgp 109
redistribute ospf...

下面的例子使得指定的 RIP 中的路由被重新分布到 OSPF 域中。

router ospf 109
redistribute rip

下面的例子中网络 20.0.0.0 在 OSPF 1 呈现为代价为 100 的外部链路状态宣告。

interface ethernet 0
ip address 20.0.0.1 255.0.0.0
ip ospf cost 100
interface ethernet 1
ip address 10.0.0.1 255.0.0.0
!
router ospf 1
network 10.0.0.0 0.255.255.255 area 0

redistribute ospf 2
router ospf 2
network 20.0.0.0 0.255.255.255 area 0

实验十 单区域 OSPF 基本配置

一、实验目的

1. 掌握单区域 OSPF 的配置。
2. 理解链路状态路由协议的工作过程。
3. 掌握实验环境中虚拟接口的配置。

二、应用环境

1. 在大规模网络中，OSPF 作为链路状态路由协议的代表应用非常广泛。
2. 具有无自环，收敛快的特点。

三、实验设备

1. DCR-1751   两台。
2. CR-V35MT   一条。
3. CR-V35FC   一条。

四、实验拓扑

该实验拓扑结构如图 2-20 所示。

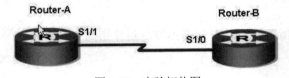

图 2-20 实验拓扑图

五、实验要求

该实验中各设备接口配置如表 2-12 所示。

表 2-12 配置表

Router-A		Router-B	
S1/1	192.168.5.1/24	S1/0	192.168.5.2/24
Loopback0	10.10.10.1/24	Loopback0	10.10.11.1/24

六、实验步骤

第一步：路由器环回接口的配置(其他接口配置请参见实验三)。

路由器 A：

Router-A_config#interface   loopback0
Router-A_config_l0#ip address 10.10.10.1 255.255.255.0

路由器 B：

Router-B#config
Router-B_config#interface loopback0
Router-B_config_l0#ip address   10.10.11.1 255.255.255.0

第二步：验证接口配置。

Router-B#sh interface loopback0
Loopback0 is up, line protocol is up
    Hardware is Loopback
    Interface address is 10.10.11.1/24
    MTU 1514 bytes, BW 8000000 kbit, DLY 500 usec
    Encapsulation LOOPBACK

第三步：路由器的 OSPF 配置。

A 的配置：

Router-A_config#router ospf 2   ！启动 OSPF 进程，进程号为 2
Router-A_config_ospf_1#network 10.10.10.0 255.255.255.0 area 0   ！注意要写掩码和区域号
Router-A_config_ospf_1#network 192.168.5.0 255.255.255.0 area 0

B 的配置：

Router-B_config#router ospf 1
Router-B_config_ospf_1#network 10.10.11.0 255.255.255.0 area 0
Router-B_config_ospf_1#network 192.168.5.0 255.255.255.0 area 0

第四步：查看路由表。

路由器 A：

Router-A#sh ip route
Codes: C - connected, S - static, R - RIP, B - BGP, BC - BGP connected
       D - DEIGRP, DEX - external DEIGRP, O - OSPF, OIA - OSPF inter area
       ON1 - OSPF NSSA external type 1, ON2 - OSPF NSSA external type 2
       OE1 - OSPF external type 1, OE2 - OSPF external type 2
       DHCP - DHCP type

VRF ID: 0

C	10.10.10.0/24	is directly connected, Loopback0
O	10.10.11.1/32	[110,1600] via 192.168.5.2(on Serial1/1)

！注意到环回接口产生的是主机路由

C	192.168.5.0/24	is directly connected, Serial1/1

路由器 B：

Router-B#show ip route

Codes: C - connected, S - static, R - RIP, B - BGP, BC - BGP connected
　　　　D - DEIGRP, DEX - external DEIGRP, O - OSPF, OIA - OSPF inter area
　　　　ON1 - OSPF NSSA external type 1, ON2 - OSPF NSSA external type 2
　　　　OE1 - OSPF external type 1, OE2 - OSPF external type 2
　　　　DHCP - DHCP type

VRF ID: 0

O	10.10.10.1/32	[110,1601] via 192.168.5.1(on Serial1/0)	！注意管理距离为 110
C	10.10.11.0/24	is directly connected, Loopback0	
C	192.168.5.0/24	is directly connected, Serial1/0	

第五步：其他验证命令。

Router-B#sh ip ospf 1　　　　　　　　　　　　　！显示该 OSPF 进程的信息
OSPF process: 1, Router ID: 192.168.2.1
Distance: intra-area 110,　inter-area 110,　external 150
SPF schedule delay 5 secs, Hold time between two SPFs 10 secs
SPFTV:11(1), TOs:24, SCHDs:27
All Rtrs support Demand-Circuit.
Number of areas is 1
AREA: 0
　　Number of interface in this area is 2(UP: 3)
　　Area authentication type:　　None
　　All Rtrs in this area support Demand-Circuit.

Router-A#show ip ospf interace　　　　　　　　！显示 OSPF 接口状态和类型
Serial1/1 is up, line protocol is up
　　Internet Address: 192.168.5.1/24
　　Nettype: Point-to-Point
　　OSPF process is 2,　AREA: 0, Router ID: 192.168.5.1
　　Cost: 1600, Transmit Delay is 1 sec, Priority 1
　　Hello interval is 10, Dead timer is 40, Retransmit is 5
　　OSPF INTF State is IPOINT_TO_POINT
　　Neighbor Count is 1, Adjacent neighbor count is 1
　　　　Adjacent with neighbor 192.168.5.2

Loopback0 is up, line protocol is up
   Internet Address: 10.10.10.1/24
   Nettype: Broadcast   ! 环回接口的网络类型默认为广播
   OSPF process is 2,   AREA: 0, Router ID: 192.168.5.1
   Cost: 1, Transmit Delay is 1 sec, Priority 1
   Hello interval is 10, Dead timer is 40, Retransmit is 5
   OSPF INTF State is ILOOPBACK
   Neighbor Count is 0, Adjacent neighbor count is 0

Router-A#sh ip ospf neighbor   ! 显示 OSPF 邻居

                            OSPF process: 2
                               AREA: 0

Neighbor ID	Pri	State	DeadTime	Neighbor Addr	Interface
192.168.2.1	1	FULL/-	31	192.168.5.2	Serial1/1

第六步：修改环回接口的网络类型。

Router-A#conf
Router-A_config#interface   loopback 0
Router-A_config_l0#ip ospf network point-to-point   ! 将类型改为点到点

第七步：查看接口状态和路由器 B 的路由表。

Router-A#sh ip ospf interface
Serial1/1 is up, line protocol is up
   Internet Address: 192.168.5.1/24
   Nettype: Point-to-Point
   OSPF process is 2,   AREA: 0, Router ID: 192.168.5.1
   Cost: 1600, Transmit Delay is 1 sec, Priority 1
   Hello interval is 10, Dead timer is 40, Retransmit is 5
   OSPF INTF State is IPOINT_TO_POINT
   Neighbor Count is 1, Adjacent neighbor count is 1
     Adjacent with neighbor 192.168.5.2

Loopback0 is up, line protocol is up
   Internet Address: 10.10.10.1/24
   Nettype: Point-to-Point
   OSPF process is 2,   AREA: 0, Router ID: 192.168.5.1
   Cost: 1, Transmit Delay is 1 sec, Priority 1
   Hello interval is 10, Dead timer is 40, Retransmit is 5
   OSPF INTF State is IPOINT_TO_POINT
   Neighbor Count is 0, Adjacent neighbor count is 0

Router-B#sh ip route
Codes: C - connected, S - static, R - RIP, B - BGP, BC - BGP connected

       D - DEIGRP, DEX - external DEIGRP, O - OSPF, OIA - OSPF inter area
       ON1 - OSPF NSSA external type 1, ON2 - OSPF NSSA external type 2
       OE1 - OSPF external type 1, OE2 - OSPF external type 2
       DHCP - DHCP type

VRF ID: 0

O   10.10.10.0/24   [110,1600] via 192.168.5.1(on Serial1/0)
C   10.10.11.0/24   is directly connected, Loopback0
C   192.168.5.0/24   is directly connected, Serial1/0

七、注意事项和排错

1. 每个路由器的 OSPF 进程号可以不同，一个路由器可以有多个 OSPF 进程。
2. OSPF 是无类路由协议，一定要加掩码。
3. 第一个区域必须是区域 0。

八、配置序列

```
Router-A#show running-conf
Building configuration...

Current configuration:
!
!version 1.3.2E
service timestamps log date
service timestamps debug date
no service password-encryption
!
hostname Router-A
!
!
interface Loopback0
  ip address 10.10.10.1 255.255.255.0
  no ip directed-broadcast
  ip ospf network point-to-point
!
interface FastEthernet0/0
  ip address 192.168.0.1 255.255.255.0
  no ip directed-broadcast
  shutdown
!
interface Serial1/0
  no ip address
```

```
    no ip directed-broadcast
    physical-layer speed 64000
   !
   interface Serial1/1
    ip address 192.168.5.1 255.255.255.0
    no ip directed-broadcast
    physical-layer speed 64000
   !
   interface Async0/0
    no ip address
    no ip directed-broadcast
   !
   !
   router ospf 2
    network 192.168.5.0 255.255.255.0 area 0
    network 10.10.10.0 255.255.255.0 area 0
   !
   !
```

九、思考题

1. OSPF 与 RIP 有哪些区别？
2. 环回接口有什么好处？

十、课后练习

请将地址改为 10.0.0.0/24 重复以上实验。

十一、相关命令详解

router ospf

配置路由器使用 OSPF 路由。no router ospf 禁止路由器使用 OSPF。

router ospf process-id

no router ospf process-id

参数：实验参数如表 2-13 所示。

表 2-13 参数说明

参数	参数说明
process-id	用于内部标示 OSPF 路由处理的参数，它是本地分配的非负整数。它唯一表示一个 OSPF 的路由处理过程

默认：无 OSPF 路由处理被定义。

命令模式：全局配置态。

使用说明：在一个路由器中，可以有多个 OSPF 路由处理过程。

举例：下面配置了一个 OSPF 路由处理，其处理号为 109。

router ospf 109

实验十一　RIP-2 邻居认证配置

一、实验目的

1. 掌握邻居认证的配置。
2. 理解 RIP-2 与 RIP-1 的不同。

二、应用环境

为避免外部路由器得到路由表，保证一定的安全性，采用认证。

三、实验设备

1. DCR-1751　　两台。
2. CR-V35MT　　一条。
3. CR-V35FC　　一条。

四、实验拓扑

该实验拓扑结构如图 2-21 所示。

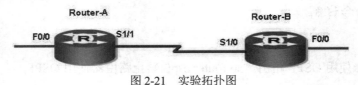

图 2-21　实验拓扑图

五、实验要求

该实验中各项配置如表 2-14 所示。

表 2-14　配置表

Router-A		Router-B	
S1/1(DCE)	192.168.5.1/24	S1/0(DTE)	192.168.5.2/24
F0/0	192.168.0.1	F0/0	192.169.2.1/24

六、实验步骤

第一步：按照实验三和表，配置路由器的所有接口地址并测试连通性。
第二步：配置路由器 B。

Router-B#conf
Router-B_config#router rip
Router-B_config_rip#version 2   ！配置为版本 2
Router-B_config_rip#network 192.168.5.0
Router-B_config_rip#network 192.168.2.0
Router-B_config_rip#exit
Router-B_config#int s1/0   ！进入与 A 相连的接口
Router-B_config_s1/0#ip rip authentication simple   ！配置以明文方式验证
Router-B_config_s1/0#ip rip password digitalchina   ！配置密码为 digitalchina
Router-B_config_s1/0#^Z

第三步：查看 B 的配置。

Router-B#sh run
Building configuration...

Current configuration:
!
!version 1.3.2E
service timestamps log date
service timestamps debug date
no service password-encryption
!
hostname Router-B
!
!
interface FastEthernet0/0
  ip address 192.168.2.1 255.255.255.0
  no ip directed-broadcast
!
interface Serial1/0
  ip address 192.168.5.2 255.255.255.0
  no ip directed-broadcast
  ip rip authentication simple
  ip rip password digitalchina
!
interface Async0/0
  no ip address
  no ip directed-broadcast
!

```
!
router rip
 version 2
 network 192.168.2.0
 network 192.168.5.0
!
!
```

第四步：配置路由器 A(不配认证)并查看路由表。

```
Router-A#conf
Router-A_config#router rip
Router-A_config_rip#version 2
Router-A_config_rip#network 192.168.0.0
Router-A_config_rip#network 192.168.5.0
Router-A_config_rip#^Z
```

查看路由表

```
Router-A#sh ip route
Codes: C - connected, S - static, R - RIP, B - BGP, BC - BGP connected
       D - DEIGRP, DEX - external DEIGRP, O - OSPF, OIA - OSPF inter area
       ON1 - OSPF NSSA external type 1, ON2 - OSPF NSSA external type 2
       OE1 - OSPF external type 1, OE2 - OSPF external type 2
       DHCP - DHCP type

VRF ID: 0

C       192.168.0.0/24       is directly connected, FastEthernet0/0    ! 没有学习到路由
C       192.168.5.0/24       is directly connected, Serial1/1
```

第五步：配置 A 的认证。

```
Router-A_config#int s1/1                              ! 进入与 B 相连的接口
Router-A_config_s1/1#ip rip authentication simple
Router-A_config_s1/1#ip rip password digitalchina
```

第六步：再次查看路由表。

```
Router-A#sh ip route
Codes: C - connected, S - static, R - RIP, B - BGP, BC - BGP connected
       D - DEIGRP, DEX - external DEIGRP, O - OSPF, OIA - OSPF inter area
       ON1 - OSPF NSSA external type 1, ON2 - OSPF NSSA external type 2
       OE1 - OSPF external type 1, OE2 - OSPF external type 2
       DHCP - DHCP type
```

VRF ID: 0

C	192.168.0.0/24	is directly connected, FastEthernet0/0
C	192.168.5.0/24	is directly connected, Serial1/1
R	192.168.2.0/24	[120,1] via 192.168.5.2(on Serial1/1)

Router-B#show ip route
Codes: C - connected, S - static, R - RIP, B - BGP, BC - BGP connected
      D - DEIGRP, DEX - external DEIGRP, O - OSPF, OIA - OSPF inter area
      ON1 - OSPF NSSA external type 1, ON2 - OSPF NSSA external type 2
      OE1 - OSPF external type 1, OE2 - OSPF external type 2
      DHCP - DHCP type

VRF ID: 0

R	192.168.0.0/24	[120,1] via 192.168.5.1(on Serial1/0)
C	192.168.5.0/24	is directly connected, Serial1/0
C	192.168.2.0/24	is directly connected, FastEthernet0/0

七、注意事项和排错

1. 只有 RIP-2 才支持认证。
2. 在相邻的接口上配置认证。
3. 认证密码要一致，必须是双向的。

八、配置序列

Router-B#sh run
Building configuration...

Current configuration:
!
!version 1.3.2E
service timestamps log date
service timestamps debug date
no service password-encryption
!
hostname Router-B
!
!
!
interface FastEthernet0/0
 ip address 192.168.2.1 255.255.255.0
 no ip directed-broadcast

```
!
interface Serial1/0
  ip address 192.168.5.2 255.255.255.0
  no ip directed-broadcast
  ip rip authentication simple
  ip rip password digital
!
interface Async0/0
  no ip address
  no ip directed-broadcast
!
!
router rip
  version 2
  network 192.168.2.0
  network 192.168.5.0
!
```

九、思考题

1. RIP-2 认证有什么意义？
2. 为什么 RIP-2 认证一定要是双向的？

十、相关命令详解

ip rip authentication

使用 ip rip authentication 接口配置命令指定用于 RIP-2 包的认证类型，no ip rip authentication 则不对报文进行认证。

ip rip authentication {simple | message-digest}

no ip rip authentication

参数：实验参数如表 2-15 所示。

表 2-15　参数说明

参数	参数说明
simple	明文认证类型
message-digest	MD5 密文认证类型

默认：不认证。

命令模式：接口配置态。

使用说明：RIP-1 不支持认证。

实验十二　多区域 OSPF 配置

一、实验目的

1. 掌握多区域 OSPF 的配置。
2. 理解 OSPF 区域的意义。

二、应用环境

在大规模网络中,我们通常划分区域减少资源消耗,并将拓扑的变化本地化。

三、实验设备

1. DCR-1751　　3 台。
2. CR-V35MT　　一条。
3. CR-V35FC　　一条。

四、实验拓扑

该实验拓扑结构如图 2-22 所示。

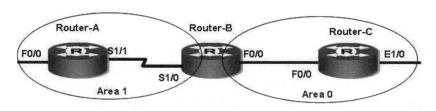

图 2-22　实验拓扑图

五、实验要求

该实验中各项配置如表 2-16 所示。

表 2-16　配置表

Router-A		Router-B		Router-C	
S1/1(DCE)	192.168.5.1	S1/0(DTE)	192.168.5.2	F0/0	192.168.2.2
F0/0	192.168.0.1	F0/0	192.168.2.1	E1/0	192.168.3.1

其中路由器 B 为 ABR。

六、实验步骤

第一步:参照实验三和表 2-16 配置各接口地址,并测试连通性。

第二步：路由器 A 的配置。

Router-A#conf
Router-A_config#router ospf 100
Router-A_config_ospf_100#network 192.168.0.0 255.255.255.0 area 1
Router-A_config_ospf_100#network 192.168.5.0 255.255.255.0 area 1
Router-A_config_ospf_100#^Z

第三步：路由器 B 的配置。

Router-B#conf
Router-B_config#router ospf 100
Router-B_config_ospf_100#network 192.168.5.0 255.255.255.0 area 1　　！注意区域的划分在接口上
Router-B_config_ospf_100#network 192.168.2.0 255.255.255.0 area 0
Router-B_config_ospf_100#^Z

第四步：路由器 C 的配置。

Router-C#conf
Router-C_config#router ospf 100
Router-C_config_ospf_100#network 192.168.2.0 255.255.255.0 area 0
Router-C_config_ospf_100#network 192.168.3.0 255.255.255.0 area 0
Router-C_config_ospf_100#^Z

第五步：查看路由表。

Router-A#sh ip route
Codes: C - connected, S - static, R - RIP, B - BGP, BC - BGP connected
　　　　D - DEIGRP, DEX - external DEIGRP, O - OSPF, OIA - OSPF inter area
　　　　ON1 - OSPF NSSA external type 1, ON2 - OSPF NSSA external type 2
　　　　OE1 - OSPF external type 1, OE2 - OSPF external type 2
　　　　DHCP - DHCP type

VRF ID: 0

C　　　192.168.0.0/24　　　　is directly connected, FastEthernet0/0
C　　　192.168.5.0/24　　　　is directly connected, Serial1/1
O IA　192.168.2.0/24　　　　[110,1601] via 192.168.5.2(on Serial1/1)
O IA　192.168.3.0/24　　　　[110,1611] via 192.168.5.2(on Serial1/1)　　！区域间的路由

Router-B#sh ip route
Codes: C - connected, S - static, R - RIP, B - BGP, BC - BGP connected
　　　　D - DEIGRP, DEX - external DEIGRP, O - OSPF, OIA - OSPF inter area
　　　　ON1 - OSPF NSSA external type 1, ON2 - OSPF NSSA external type 2
　　　　OE1 - OSPF external type 1, OE2 - OSPF external type 2
　　　　DHCP - DHCP type

VRF ID: 0

O	192.168.0.0/24	[110,1601] via 192.168.5.1(on Serial1/0)
C	192.168.5.0/24	is directly connected, Serial1/0
C	192.168.2.0/24	is directly connected, FastEthernet0/0
O	192.168.3.0/24	[110,11] via 192.168.2.2(on FastEthernet0/0)

！对 ABR 来说是区域内的路由

Router-C#sh ip route
Codes: C - connected, S - static, R - RIP, B - BGP
        D - DEIGRP, DEX - external DEIGRP, O - OSPF, OIA - OSPF inter area
        ON1 - OSPF NSSA external type 1, ON2 - OSPF NSSA external type 2
        OE1 - OSPF external type 1, OE2 - OSPF external type 2

O IA 192.168.0.0/24   [110,1602] via 192.168.2.1(on   FastEthernet0/0)
O IA 192.168.5.1/32   [110,1601] via 192.168.2.1(on   FastEthernet0/0)
O IA 192.168.5.2/32   [110,3201] via 192.168.2.1(on   FastEthernet0/0)   ！区域间的路由
C   192.168.2.0/24   is directly connected,   FastEthernet0/0
C   192.168.3.0/24   is directly connected,   Ethernet1/0

七、注意事项和排错

1. 区域的划分在接口上进行。

2. 必须有 area 0 存在。

八、配置序列

Router-B#sh run
Building configuration...

Current configuration:
!
!version 1.3.2E
service timestamps log date
service timestamps debug date
no service password-encryption
!
hostname Router-B
!
ip host a 192.168.5.1
ip host c 192.168.2.2
!
!
!
!

```
!
!
interface FastEthernet0/0
  ip address 192.168.2.1 255.255.255.0
  no ip directed-broadcast
!
interface Serial1/0
  ip address 192.168.5.2 255.255.255.0
  no ip directed-broadcast
!
interface Async0/0
  no ip address
  no ip directed-broadcast
!
!
!
router ospf 100
  network 192.168.5.0 255.255.255.0 area 1
  network 192.168.2.0 255.255.255.0 area 0
!
!
!
```

九、思考题

1. 为什么必须有 area 0 存在？
2. 在路由器 A 和 C 宣告网段时有其他方法吗？

十、相关命令详解

network area

将一个区域中几个网段定义成一个网络范围，no network area 命令取消网络范围。

network network mask area area_id [advertise | not-advertise]

no network network mask area area_id [advertise | not-advertise]

参数：实验参数如表 2-17 所示。

表 2-17　参数说明

参数	参数说明
network	网络 IP 地址，点分十进制格式
mask	掩码，点分十进制格式
area_id	为区域号
advertise 和 notadvertise	指定是否将到这一网络范围路由的摘要信息广播出去

默认：系统默认没有配置网络范围。

命令模式：路由配置态。

使用说明：一旦将某一网络的范围加入到区域中，到区域中所有落在这一范围内的 IP 地址的内部路由都不再被独立地广播到别的区域，而只是广播整个网络范围路由的摘要信息。引入网络范围和对该范围的限定，可以减少区域间路由信息的交流量。

举例：定义网络范围 10.0.0.0 255.0.0.0 加入到区域 2 中。

router_config_ospf_10#network 10.0.0.0 255.0.0.0 area 2

实训十一　多台路由器间的动态 OSPF 路由配置

一、实训设备

(1) DCS-3926S 交换机　　2 台。
(2) DCR-1702E 路由器　　1 台；DCR-2631 路由器　　1 台。
(3) PC 机　　4 台。
(4) 直通网线　　7 根。

二、实训拓扑

该实验拓扑结构如图 2-23 所示。

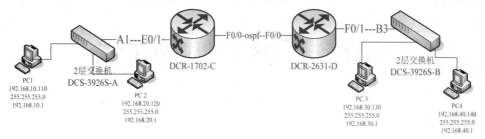

图 2-23　网络拓扑结构

三、实训要求及步骤

具体要求：按照图 2-23 所示进行连线，各个设备做如下配置。

(1) 交换机 A 划分 vl 10(9-12)，vl 20(13-16)，Trunk (1)；
(2) 交换机 B 划分 vl 30(17-20)，vl 40(21-24)，Trunk (3)。
(3) 设置路由器 C：

E 0/1.10　　192.168.10.1/24　　封装 vl 10；
E 0/1.20　　192.168.20.1/24　　封装 vl 20；
F 0/0　　　192.168.100.1/30

(4) 设置路由器 D：

F 0/1.30   192.168.30.1/24   封装 vl 30；
F 0/1.40   192.168.40.1/24   封装 vl 40；
F 0/0   192.168.100.2/30

(5) 启用路由器的动态 OSPF 协议，配置 C 的动态路由 OSPF 协议：

RC_config#router ospf 1
RC_config_Router#network 192.168.10.0 255.255.255.0 area 0
RC_config_Router#network 192.168.20.0 255.255.255.0 area 0
RC_config_Router#network 192.168.100.0 255.255.255.252 area 0

(6) 配置 D 的动态路由 OSPF 协议：

RD_config#router ospf 2
RD_config_Router#network 192.168.30.0 255.255.255.0 area 0
RD_config_Router#network 192.168.40.0 255.255.255.0 area 0
RD_config_Router#network 192.168.100.0 255.255.255.252 area 0

验证：sh ip ospf   sh ip route

(7) 根据连线位置正确配置 PC1、PC2 的地址(注意配上网关地址以及与 VLAN 的对应关系)。

实验结果：PC1---ping---PC2、PC3、PC4   通。
查看设备状态：show vlan   show run   show ip route。

实验十三 OSPF 邻居认证配置

一、实验目的

1. 掌握 OSPF 邻居认证的配置。
2. 理解理解邻居认证的作用。

二、应用环境

在企业环境中，需要配置认证来保证 OSPF 路由的安全性。

三、实验设备

1. DCR-1751   两台。
2. CR-V35MT   一条。
3. CR-V35FC   一条。

四、实验拓扑

该实验拓扑结构如图 2-24 所示。

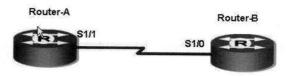

图 2-24　实验拓扑图

五、实验要求

该实验中各项配置如表 2-18 所示。

表 2-18　配置表

Router-A		Router-B	
S1/1	192.168.5.1/24	S1/0	192.168.5.2/24
Loopback0	10.10.10.1/24	Loopback0	10.10.11.1/24

六、实验步骤

第一步：路由器环回接口的配置(其他接口配置请参见实验三)。

路由器 A：

Router-A_config#interface　loopback0
Router-A_config_l0#ip address 10.10.10.1 255.255.255.0

路由器 B：

Router-B#config
Router-B_config#interface loopback0
Router-B_config_l0#ip address　10.10.11.1 255.255.255.0

第二步：验证接口配置。

Router-B#sh interface loopback0
Loopback0 is up, line protocol is up
　　Hardware is Loopback
　　Interface address is 10.10.11.1/24
　　MTU 1514 bytes, BW 8000000 kbit, DLY 500 usec
　　Encapsulation LOOPBACK

第三步：路由器的 OSPF 配置。
A 的配置：

Router-A_config#router ospf 2　　　　　　　　　　　　　　　！启动 OSPF 进程，进程号为 2

```
Router-A_config_ospf_1#network 10.10.10.0 255.255.255.0 area 0    ！注意要写掩码和区域号
Router-A_config_ospf_1#network 192.168.5.0 255.255.255.0 area 0
Router-A_config_ospf_1#area 0 authen simple              ！定义在区域 0 中使用明文认证
Router-A_config_ospf_1#exit
Router-A_config#interface s1/1
Router-A_config#ip ospf password digitalchina             ！配置接口密码
```

B 的配置：

```
Router-B_config#router ospf 1
Router-B_config_ospf_1#network 10.10.11.0 255.255.255.0 area 0
Router-B_config_ospf_1#network 192.168.5.0 255.255.255.0 area 0
Router-B_config_ospf_1# area 0 authen simple             ！定义在区域 0 中使用明文认证
Router-B_config_ospf_1#exit
Router-B_config#interface s1/0
Router-B_config#ip ospf password digitalchina            ！配置接口密码
```

第四步：查看路由表。

路由器 A：

```
Router-A#sh ip route
Codes: C - connected, S - static, R - RIP, B - BGP, BC - BGP connected
       D - DEIGRP, DEX - external DEIGRP, O - OSPF, OIA - OSPF inter area
       ON1 - OSPF NSSA external type 1, ON2 - OSPF NSSA external type 2
       OE1 - OSPF external type 1, OE2 - OSPF external type 2
       DHCP - DHCP type

VRF ID: 0

C       10.10.10.0/24        is directly connected, Loopback0
O       10.10.11.1/32        [110,1600] via 192.168.5.2(on Serial1/1)
                                              ！注意到环回接口产生的是主机路由
C       192.168.5.0/24       is directly connected, Serial1/1
```

路由器 B：

```
Router-B#show ip route
Codes: C - connected, S - static, R - RIP, B - BGP, BC - BGP connected
       D - DEIGRP, DEX - external DEIGRP, O - OSPF, OIA - OSPF inter area
       ON1 - OSPF NSSA external type 1, ON2 - OSPF NSSA external type 2
       OE1 - OSPF external type 1, OE2 - OSPF external type 2
       DHCP - DHCP type

VRF ID: 0

O       10.10.10.1/32        [110,1601] via 192.168.5.1(on Serial1/0) ！注意管理距离为 110
```

C	10.10.11.0/24	is directly connected, Loopback0
C	192.168.5.0/24	is directly connected, Serial1/0

七、注意事项和排错

1. 在邻居接口上配置认证。
2. 认证方式除了明文，还有密钥方式。

八、配置序列

无。

九、思考题

1. 认证的作用是什么？
2. 在什么地方配置认证？

十、课后练习

请将地址改为 10.0.0.0/24 重复以上实验。

十一、相关命令详解

ip ospf password

为邻接路由配置态口令。使用 no ip ospf password 取消设置。

ip ospf password password

no ip ospf password

参数：实验参数如表 2-19 所示。

表 2-19 参数说明

参数	参数说明
password	任何连续的 8 位字符串

默认：无口令。

命令模式：接口配置态。

使用说明：这个命令生成的口令直接插入 OSPF 路由信息包。可以为每个接口的每个网络配置一个口令。所有的邻居路由器必须有相同的口令才能交换 OSPF 路由信息。

注意：这个命令仅在通过命令 area authentucation 设置允许认证才生效。

举例：

ip ospf password yourpass

实验十四 OSPF 路由汇总配置

一、实验目的

1. 掌握 OSPF 的配置。
2. 理解 OSPF 路由汇总的意义。

二、应用环境

在大规模网络中，路由表非常庞大，降低了转发速度，通常在子网边界做汇总，这样可以减少路由表的长度。

三、实验设备

1. DCR-1751　　三台。
2. CR-V35MT　　一条。
3. CR-V35FC　　一条。

四、实验拓扑

该实验拓扑结构如图 2-25 所示。

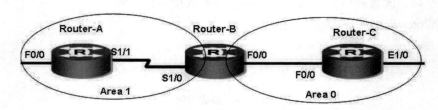

图 2-25　实验拓扑图

五、实验要求

该实验中各设备接口配置如表 2-20 所示。

表 2-20　配置表

Router-A		Router-B		Router-C	
S1/1(DCE)	10.10.11.1/24	S1/0(DTE)	10.10.11.2/24	F0/0	192.168.2.2/24
F0/0	10.10.10.1/24	F0/0	192.168.2.1/24	E1/0	192.168.3.1/24

其中路由器 B 为 ABR。

六、实验步骤

第一步：参照实验三和表 2-20 配置各接口地址，并测试连通性。

第二步：路由器 A 的配置。

Router-A#conf
Router-A_config#router ospf 100
Router-A_config_ospf_100#network 10.10.10.0 255.255.255.0 area 1
Router-A_config_ospf_100#network 10.10.11.0 255.255.255.0 area 1
Router-A_config_ospf_100#^Z

第三步：路由器 B 的配置。

Router-B#conf
Router-B_config#router ospf 100
Router-B_config_ospf_100#network 10.10.11.0 255.255.255.0 area 1
！注意区域的划分在接口上
Router-B_config_ospf_100#network 192.168.2.0 255.255.255.0 area 0
Router-B_config_ospf_100#^Z

第四步：路由器 C 的配置。

Router-C#conf
Router-C_config#router ospf 100
Router-C_config_ospf_100#network 192.168.2.0 255.255.255.0 area 0
Router-C_config_ospf_100#network 192.168.3.0 255.255.255.0 area 0
Router-C_config_ospf_100#^Z

第五步：查看路由表。

Router-A#sh ip route
Codes: C - connected, S - static, R - RIP, B - BGP, BC - BGP connected
   D - DEIGRP, DEX - external DEIGRP, O - OSPF, OIA - OSPF inter area
   ON1 - OSPF NSSA external type 1, ON2 - OSPF NSSA external type 2
   OE1 - OSPF external type 1, OE2 - OSPF external type 2
   DHCP - DHCP type

VRF ID: 0

C	10.10.10.0/24	is directly connected, FastEthernet0/0	
C	10.10.11.0/24	is directly connected, Serial1/1	
O IA	192.168.2.0/24	[110,1601] via 192.168.5.2(on Serial1/1)	
O IA	192.168.3.0/24	[110,1611] via 192.168.5.2(on Serial1/1)	！区域间的路由

Router-B#sh ip route
Codes: C - connected, S - static, R - RIP, B - BGP, BC - BGP connected
   D - DEIGRP, DEX - external DEIGRP, O - OSPF, OIA - OSPF inter area

```
        ON1 - OSPF NSSA external type 1, ON2 - OSPF NSSA external type 2
        OE1 - OSPF external type 1, OE2 - OSPF external type 2
        DHCP - DHCP type

VRF ID: 0

O       10.10.10.0/24        [110,1601] via 192.168.5.1(on Serial1/0)
C       10.10.11.0/24        is directly connected, Serial1/0
C       192.168.2.0/24       is directly connected, FastEthernet0/0
O       192.168.3.0/24       [110,11] via 192.168.2.2(on FastEthernet0/0)
                                   ！对 ABR 来说是区域内的路由


Router-C#sh ip route
Codes: C - connected, S - static, R - RIP, B - BGP
       D - DEIGRP, DEX - external DEIGRP, O - OSPF, OIA - OSPF inter area
       ON1 - OSPF NSSA external type 1, ON2 - OSPF NSSA external type 2
       OE1 - OSPF external type 1, OE2 - OSPF external type 2

O IA 10.10.10.0/24     [110,1602] via 192.168.2.1(on  FastEthernet0/0)
O IA 10.10.11.1/32     [110,1601] via 192.168.2.1(on  FastEthernet0/0)
O IA 10.10.11.2/32     [110,3201] via 192.168.2.1(on  FastEthernet0/0)    ！区域间的路由
C    192.168.2.0/24    is directly connected,   FastEthernet0/0
C    192.168.3.0/24    is directly connected,   Ethernet1/0
```

第六步：在路由器 B 上做路由汇总。

```
Router-B#conf
Router-B_config#router ospf 100
Router-B_config_ospf_100#area 1 range 10.10.0.0 255.255.0.0
Router-B_config_ospf_100#network 192.168.2.0 255.255.255.0 area 0
Router-B_config_ospf_100#^Z
```

第七步：再次查看路由器 C 上的路由表。

```
Router-C#sh ip route
Codes: C - connected, S - static, R - RIP, B - BGP
       D - DEIGRP, DEX - external DEIGRP, O - OSPF, OIA - OSPF inter area
       ON1 - OSPF NSSA external type 1, ON2 - OSPF NSSA external type 2
       OE1 - OSPF external type 1, OE2 - OSPF external type 2

O IA 10.10.0.0/16      [110,1602] via 192.168.2.1(on  FastEthernet0/0)   ！注意新的掩码
C    192.168.2.0/24    is directly connected,   FastEthernet0/0
C    192.168.3.0/24    is directly connected,   Ethernet1/0
```

七、注意事项和排错

1. 实际环境中，通常做精确的汇总。
2. 汇总操作在边界路由器上进行。

八、配置序列

```
Router-B#sh run
Building configuration...

Current configuration:
!
!version 1.3.2E
service timestamps log date
service timestamps debug date
no service password-encryption
!
hostname Router-B
!
!
interface FastEthernet0/0
  ip address 192.168.2.1 255.255.255.0
  no ip directed-broadcast
!
interface Serial1/0
  ip address 10.10.11.2 255.255.255.0
  no ip directed-broadcast
!
interface Async0/0
  no ip address
  no ip directed-broadcast
!
!
!
router ospf 100
  area 1 range 10.10.0.0 255.255.0.0
  network 192.168.2.0 255.255.255.0 area 0
!
!
!
```

九、思考题

1. 路由汇总的作用是什么？
2. 汇总操作通常在什么地方进行？

十、课后练习

请将地址改为 10.0.0.0/25 重复以上实验。

十一、相关命令详解

area range

在域边界进行路由汇总。用 no area range 取消设置。

area area-id range address mask[not-advertise]

no area area-id range address mask not-advertise

no area area-id range address mask

no area area-id

参数：实验参数如表 2-21 所示。

表 2-21 参数说明

参数	参数说明
password	表示要进行路由汇总的域。可以是十进制数，也可以是一个 IP 地址
address	IP 地址
mask	IP 掩码
advertise	汇总后发布
not-advertise	汇总后不发布

默认：不起作用。

命令模式：路由配置态。

使用说明：area range 命令仅仅用在 ABR 路由器上。作用是 ABR 使用一条汇总路由广播到其他路由器。这样在域边界路由被缩小，对于区域外部，每一个地址范围只有唯一一条汇总路由，这就是路由汇总。

这个命令可以在多个区的路由器上进行配置，这样 OSPF 能汇总多个地址范围。

注意：使用命令 no area area-id(无其他参数)取消设置时，它取消所有的域参数子命令，如：area authentication、area default-cost、area nssa、area range、area stub 与 area virtual-link。

举例：下面的例子配置了 ABR 路由器对于子网 36.0.0.0 和所有 192.42.110.0 的主机的汇总路由。

```
interface ethernet 0
ip address 192.42.110.201 255.255.255.0
!
interface ethernet 1
ip address 36.56.0.201 255.255.0.0
!
router ospf 201
network 36.0.0.0 255.0.0.0 area 36.0.0.0
network 192.42.110.0 255.0.0.0 area 0
```

area 36.0.0.0 range 36.0.0.0 255.0.0.0
area 0 range 192.42.110.0 255.255.255.0

实验十五　NAT 地址转换的配置

一、实验目的

1. 掌握地址转换的配置。
2. 掌握向外发布内部服务器地址转换的方法。
3. 掌握私有地址访问 Internet 的配置方法。

二、应用环境

1. 企业内部有对 Internet 提供服务的 Web 服务器。
2. 企业内部使用私有地址的主机需要访问 Internet。

三、实验设备

1. DCR-1751　两台。
2. PC 机　两台。

四、实验拓扑

该实验拓扑结构如图 2-26 所示。

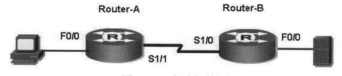

图 2-26　实验拓扑图

五、实验要求

该实验中各项配置按表 2-22 所示。

表 2-22　配置表

Router-A		Router-B		PC		Server	
S1/1(DCE)	192.168.5.1/24	S1/0	10.10.11.2/24	IP	192.168.0.3/24	IP	192.168.2.2/24
F0/0	192.168.0.1/24	F0/0	192.168.2.1/24	网关	192.168.0.1	网关	192.168.2.1

六、实验步骤

内部的 PC 需要访问外部的服务器。

假设在 Router-A 上做地址转换,将 192.168.0.0/24 转换成 192.168.5.10~192.168.5.20 之间的地址,并且做端口的地址复用。

第一步:按实验三和表 2-22 所示将接口地址和 PC 地址配置好,并且做连通性测试。

第二步:配置 Router-A 的 NAT。

```
Router-A#conf
Router-A_config#ip access-list standard 1                  !定义访问控制列表
Router-A_config_std_nacl#permit 192.168.0.0 255.255.255.0  !定义允许转换的源地址范围
Router-A_config_std_nacl#exit
Router-A_config#ip nat pool overld 192.168.5.10 192.168.5.20 255.255.255.0
!定义名为 overld 的转换地址池
Router-A_config#ip nat inside source list 1 pool overld overload
!配置将 ACL 允许的源地址转换成 overld 中的地址,并且做 PAT 的地址复用
Router-A_config#int f0/0
Router-A_config_f0/0#ip nat inside                         !定义 F0/0 为内部接口
Router-A_config_f0/0#int s1/1
Router-A_config_s1/1#ip nat outside                        !定义 S1/1 为外部接口
Router-A_config_s1/1#exit
Router-A_config#ip route 0.0.0.0 0.0.0.0 192.168.5.2       !配置路由器 A 的默认路由
```

第三步:查看 Router-B 的路由表。

```
Router-B#sh ip route
Codes: C - connected, S - static, R - RIP, B - BGP, BC - BGP connected
       D - DEIGRP, DEX - external DEIGRP, O - OSPF, OIA - OSPF inter area
       ON1 - OSPF NSSA external type 1, ON2 - OSPF NSSA external type 2
       OE1 - OSPF external type 1, OE2 - OSPF external type 2
       DHCP - DHCP type

VRF ID: 0

C       192.168.5.0/24        is directly connected, Serial1/0
C       192.168.2.0/24        is directly connected, FastEthernet0/0
```
!注意:并没有到 192.168.0.0 的路由

第四步:测试。

测试结果如图 2-27 所示。

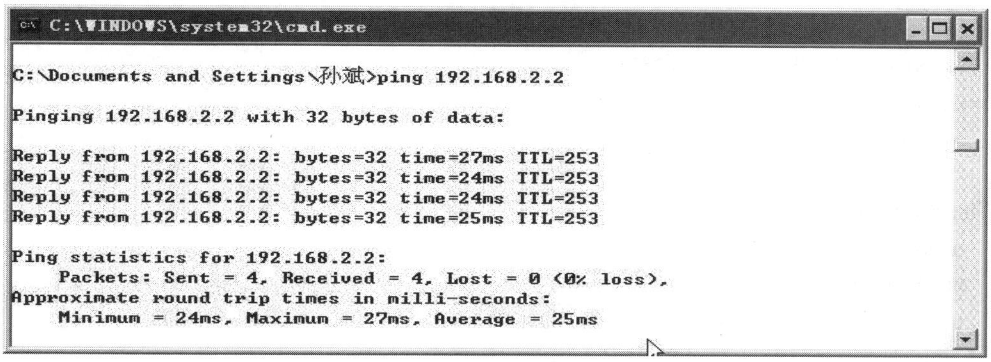

图 2-27 实验测试结果

第五步：查看地址转换表。

Router-A#sh ip nat translations
Pro. Dir   Inside local   Inside global   Outside local   Outside global

ICMP OUT 192.168.0.3:512   192.168.5.10:12512 192.168.5.2:12512   192.168.5.2:12512

注意端口的转换。

七、注意事项和排错

1. 注意转换的方向和接口。
2. 注意地址池、ACL 的名称。
3. 需要配置 A 的默认路由。

八、配置序列

Router-A#sh run
　　Building configuration...

　　Current configuration:
　　!
　　!version 1.3.2E
　　service timestamps log date
　　service timestamps debug date
　　no service password-encryption
　　!
　　hostname Router-A
　　!
　　!
　　!
　　!
　　!
　　!

```
interface FastEthernet0/0
  ip address 192.168.0.1 255.255.255.0
  no ip directed-broadcast
  ip nat inside
!
interface Serial1/0
  no ip address
  no ip directed-broadcast
  physical-layer speed 64000
!
interface Serial1/1
  ip address 192.168.5.1 255.255.255.0
  no ip directed-broadcast
  physical-layer speed 64000
  ip nat outside
!
interface Async0/0
  no ip address
  no ip directed-broadcast
!
!
!
ip route default 192.168.5.2
!
!
ip access-list standard 1
  permit 192.168.0.0 255.255.255.0
!
!
!
!
!
ip nat pool overld 192.168.5.10 192.168.5.20 255.255.255.0
ip nat inside source list 1 pool overld overload
!
```

九、思考题

1. 为什么在 B 上不需要配置 192.168.0.0 的路由就能够通信？

2. 为什么在 A 上需要配置默认路由？

3. 如果外部接口地址是通过拨号动态获得，那么该如何配置？

4. 请指出上述配置中的 Inside local、Inside global、Outside local、Outside global 分别是什么？

十、课后练习

请在 Router-B 上配置发布内部服务器 192.168.2.2/24 到外部地址 192.168.5.2/24，具体步骤如下：

(1) 在服务器 192.168.2.2/24 上安装 FTP-SERVER。
(2) 在 B 上配置静态转换。
(3) 在 PC 机上访问 192.168.5.2 的 FTP 服务。
(4) 通过 show 查看转换表。

十一、相关命令详解

ip nat inside source

使用 ip nat inside source 全局配置命令，开启内部源地址的 NAT。用这个命令的 no 形式可以删除静态翻译或删除和池的动态关联，注意：动态翻译规则和静态网段翻译规则在使用时不能删除。

动态 NAT：

ip nat inside source {list access-list-name} {interface type number | pool pool-name} [overload]
no ip nat inside source {list access-list-name} {interface type number | pool pool-name} [overload]

静态单个地址 NAT：

ip nat inside source {static {local-ip global-ip}
no ip nat inside source {static {local-ip global-ip}

静态端口 NAT：

ip nat inside source {static {tcp | udp local-ip local-port {global-ip | interface type number} global-port}
no ip nat inside source {static {tcp | udp local-ip local-port {global-ip | interface type number} global-port}

静态网段 NAT：

ip nat inside source {static {network local-network global-network mask}
no ip nat inside source {static {network local-network global-network mask}

参数：实验参数如表 2-23 所示。

表 2-23 参数说明

参数	参数说明
List access-list-name	IP 访问列表的名字。源地址符合访问列表的报文将被地址池中的全局地址来翻译
pool name	地址池的名字，从这个池中动态地分配全局 IP 地址
interface type number	指定网络接口
overload	(可选)使路由器对多个本地地址使用一个全局地址。当 overload 被设置后，相同或者不同主机的多个会话将用 TCP 或 UDP 端口号来区分

(续表)

参数	参数说明
static local-ip	建立一条独立的静态地址翻译；参数为分配给内部网主机的本地 IP 地址。这个地址可以自由选择，或从 RFC 1918 中分配
local-port	设置本地 TCP/UDP 端口号，范围为 1~65535
static global-ip	建立一条独立的静态地址翻译；这个参数为内部主机建立一个外部的网络可以唯一访问的 IP 地址
global-port	设置全局 TCP/UDP 端口号，范围为 1~65535
tcp	设置 TCP 端口翻译
udp	设置 UDP 端口翻译
network local-network	设置本地网段翻译
global-network	设置全局网段翻译
mask	设置网段翻译的网络掩码

默认：任何内部源地址的 NAT 都不存在。

命令模式：全局配置态。

使用说明：

这个命令有两种形式：动态的和静态的地址翻译。带有访问列表的格式建立动态翻译。来自和标准访问列表相匹配的地址的报文，将用指定的池中分配的全局地址来进行地址翻译，这个池是用 ip nat pool 命令所指定的。

可以作为替代方法，带有关键字 STATIC 的语法格式创建一条独立的静态地址翻译。

举例：下面的例子把来自 192.168.5.0 或 192.168.2.0 网络的内部主机间进行通信的 IP 地址翻译为 171.69.233.208/28 网络中全局唯一的 IP 地址。

```
ip nat pool net-208 171.69.233.208 171.69.233.223 255.255.255.240
ip nat inside source list a1 pool net-208
!
interface ethernet 0
ip address 171.69.232.182 255.255.255.240
ip nat outside
!
interface ethernet 1
ip address 192.168.5.94 255.255.255.0
ip nat inside
!
ip access-list standard a1
permit 192.168.5.0 255.255.255.0
permit 192.168.2.0 255.255.255.0
!
```

附 录

一、二层接入三层连通加聚合综合性实验

该实验拓扑结构如图 1 所示。

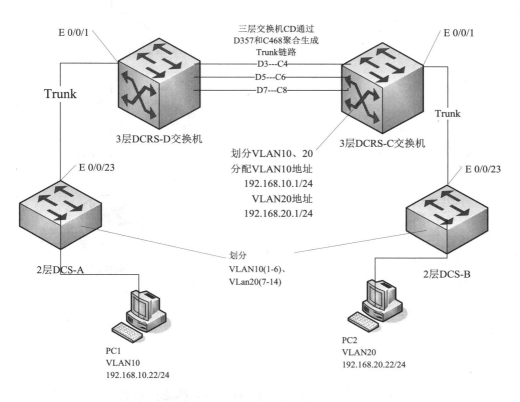

图 1　实验拓扑图

如图 1 所示，交换机 CD 间先启动生成树，经过配置，使 PC1----ping---PC2 相通，并写出详细配置。

二、单臂路由加聚合综合性实验

该实验拓扑结构如图 2 所示。

如图 2 所示，交换机 AB 间先启动生成树，经过配置，使 PC1----ping---PC2 相通，并写出详细配置。

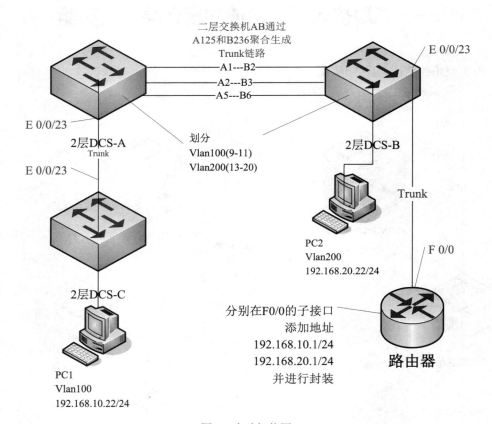

图 2　实验拓扑图

三、动态路由 RIP OSPF 综合性实验

该实验拓扑结构如图 3 所示。

如图 3 所示，交换机 AB 划分对应 VLAN，在路由器 AB 上作相应配置，并使用动态路由 RIP-2，经过配置，使 PC1----ping---PC2 相通，并写出详细配置。

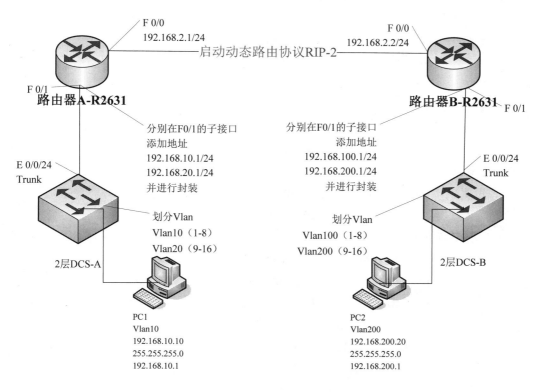

图 3　实验拓扑图

四、路由与多层交换机间的 RIP OSPF 和静态路由重分布综合性实验

该实验拓扑结构如图 4 所示。

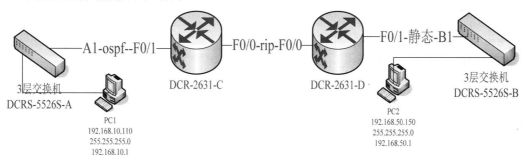

图 4　实验拓扑图

具体要求：按照图 4 所示进行连线，各个设备做如下配置。

(1) 三层交换机 A：

　　　划分　　　vl 100(9-16)，　　vl 200 (1)；
　　　　　　　　int vl100　　　　192.168.10.1/24；
　　　　　　　　int vl200　　　　192.168.20.1/30；

(2) 设置路由器 C：

　　　设置　　　　F 0/1　　192.168.20.2/30；
　　　　　　　　　F 0/0　　192.168.30.2/30；

(3) 设置路由器 D：

　　　设置　　　　F 0/0　　192.168.30.1/30；
　　　　　　　　　F 0/1　　192.168.40.2/30；

(4) 三层交换机 B：

　　　划分　　　vl 400(17-24)，　　vl 200(1)；
　　　　　　　　int vl200　　　　192.168.40.1/30；
　　　　　　　　int vl400　　　　192.168.50.1/24；

(5) PC1 设置：　　　　PC2 设置：

　　　192.168.10.110　　192.168.50.150
　　　255.255.255.0　　 255.255.255.0
　　　192.168.10.1　　　192.168.50.1

(6) 根据连线位置正确配置 PC1、PC2 的地址(注意配上网关地址以及与 VLAN 的对应关系)。

实验结果：PC1---ping---PC2　　通。

查看交换机状态：sh vlan　　　sh run　　　sh ip route。

主要配置命令参考

启动 SA 到 RC 使用动态路由 OSPF 协议，不同版本的三层交换机命令不同(方法 1 命令不能使用时，可使用方法 2)。

方法 1：

SA(config)#int vl 100
SA(config)# ip ospf enable area 0
SA(config)#int vl 200
SA(config)# ip ospf enable area 0
SA(config)#router ospf
SA(config-Router-ospf)#redistribute ospfase connected
SA(config-Router-ospf)#redistribute ospfase static
SA(config-Router-ospf)#redistribute ospfase rip

方法 2：

SA(config)#router ospf
SA(config-Router)#network 192.168.10.0 255.255.255.0 area 0
SA(config-Router)#network 192.168.20.0 255.255.255.252 area 0
SA(config-Router)#redistribute   connected
SA(config-Router)#redistribute   static
SA(config-Router)#redistribute   rip

RC_config#router ospf   1
RC_config_Router#network 192.168.20.0 255.255.255.252 area 0
RC_config_Router# redistribute connect
RC_config_Router# redistribute static
RC_config_Router# redistribute rip

启动 RC---RD 的动态路由 rip 协议

RC_config#router rip
RC_config_rip#version 2
RC_config_rip#network 192.168.30.0
RC_config_rip#auto-summary   ----自动汇总
RC_config_rip#redistribute ospf   1   ----转发 ospf 路由
RC_config_rip#redistribute connect   ----转发直连路由
RC_config_rip#redistribute static   ----转发静态路由
RD_config#router rip
RD_config_rip#version 2
RD_config_rip#network 192.168.30.0
RD_config_rip#auto-summary
RD_config_rip#redistribute connect
RD_config_rip#redistribute static

启动 RD---SB 的静态路由协议

RD_config#ip route   192.168.50.0 255.255.255.0 192.168.40.1
SB(config)#ip route   192.168.30.0 255.255.255.252 192.168.40.2
SB(config)#ip route   192.168.20.0 255.255.255.252 192.168.40.2
SB(config)#ip route   192.168.10.0 255.255.255.0 192.168.40.2

验证：sh run   sh ip route
验证命令：ping
PC1：运行 cmd

1. ping 192.168.10.1   -t   通
2. ping 192.168.20.1   -t   通
3. ping 192.168.30.1   -t   通
4. ping 192.168.40.1   -t   通

5. ping 192.168.50.1   -t   通
6. ping 192.168.50.150 -t   通

PC2：运行 cmd
1. ping 192.168.10.1   -t   通
2. ping 192.168.20.1   -t   通
3. ping 192.168.30.1   -t   通
4. ping 192.168.40.1   -t   通
5. ping 192.168.50.1   -t   通
6. ping 192.168.10.110 -t   通。

五、路由与多层交换机间的静-O-R 综合性实验

该实验拓扑结构如图 5 所示。

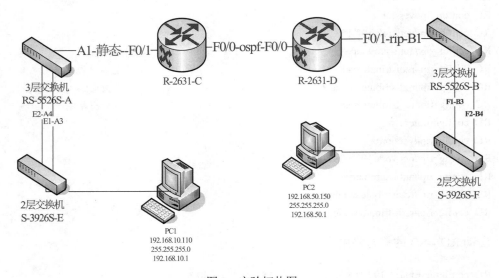

图 5　实验拓扑图

具体要求：按照图 5 所示进行连线，各个设备做如下配置。

(1) 三层交换机 A：

　　　划分　　　vl 100(9-16)，　　vl 200 (1)；trunk(3-4)
　　　　　　　int vl100　　　　192.168.10.1/24；
　　　　　　　int vl200　　　　192.168.20.1/30；

(2) 设置路由器 C：

　　　设置　　　F 0/1　　　192.168.20.2/30；
　　　　　　　　F 0/0　　　192.168.30.2/30

(3) 设置路由器 D：
　　设置　　　　F 0/0　　192.168.30.1/30
　　　　　　　　F 0/1　　192.168.40.2/30；

(4) 三层交换机 B：
　　划分　　vl 400(17-24)，　vl 200(1)；Trunk(3-4)
　　　　　　int vl200　　　　192.168.40.1/30；
　　　　　　int vl400　　　　192.168.50.1/24；

(5) 二层交换机 E：
　　划分 Vlan 100 (9-16)，vl 200(17-24)，Trunk(1-2)
　　E12 与 A34 生成聚合的 Trunk 链路。

(6) 二层交换机 F：
　　划分 Vlan 400 (9-16)，vl 200(17-24)，Trunk(1-2)
　　F12 与 B34 生成聚合的 Trunk 链路。

(7) 根据连线位置正确配置 PC1、PC2 的地址(注意配上网关地址以及与 VLAN 的对应关系)。

　　实验结果：PC1---ping---PC2　　　通。
　　查看交换机状态：sh vlan　　　sh run　　　sh ip route。

参 考 文 献

[1] 神州数码网络设备使用说明书,2008